Advanced PLC Programming

Advanced PLC Programming

Dr. Majid Pakdel

CWP

Central West Publishing

Disclaimer
Every effort has been made by the publisher and authors while preparing this book, however, no warranties are made regarding the accuracy and completeness of the content. The publisher and authors disclaim without any limitation all warranties as well as any implied warranties about sales, along with fitness of the content for a particular purpose. Citation of any website and other information sources does not mean any endorsement from the publisher and authors. For ascertaining the suitability of the contents contained herein for a particular lab or commercial use, consultation with the subject expert is needed. In addition, while using the information and methods contained herein, the practitioners and researchers need to be mindful for their own safety, along with the safety of others, including the professional parties and premises for whom they have professional responsibility. To the fullest extent of law, the publisher and authors are not liable in all circumstances (special, incidental, and consequential) for any injury and/or damage to persons and property, along with any potential loss of profit and other commercial damages due to the use of any methods, products, guidelines, procedures contained in the material herein.

NATIONAL LIBRARY OF AUSTRALIA

A catalogue record for this book is available from the National Library of Australia

ISBN (print): 978-1-925823-79-0

Contents

Preface

Programmable logic controllers (PLCs) are evolving every day and new technologies are being added to its capabilities. PLC has been introduced as an alternative to relay control systems then logical and mathematical operation functions have gradually been added to it. Today, PLCs control many of the automation processes. PLCs come in smaller sizes with faster CPUs and different network and Internet technologies. In the present book, advanced methods of PLC programming are described. Reading of this book is strongly recommended for PLC professional programmers, university professors, students, professors of technical and vocational schools, as well as those with no prior knowledge of PLC. The book is written on the assumption that the reader has no prior knowledge of PLC systems and its programming methods. The main source of information for a particular PLC is always the user manuals provided by the PLC manufacturer. This book is not a replacement for the references provided by the manufacturer, but is intended to supplement, clarify and extend this information. Despite the different types of PLCs on the market, it is not possible to give details of all the PLC makers in one book. This book is about PLCs in general and this book can be used for different types of PLC makers. Most of the instructions and applications for this book are related to Siemens PLCs (Logo and Step 7). The principles and concepts of PLC programming described in this book are shared by different PLC makers and complement the information and training programs provided by different PLC makers. Various books on PLC have been published by various authors and publishers. Unfortunately, most published books are too superficial and vacuous to express their concepts, principles, and programming. Beginners, and even those familiar with the principles of traditional relay circuits, are having trouble finding new control and designing PLC relay logic to control industrial and non-industrial processes after reading these books. They have not been understood and the books have not addressed this issue. It is easy to draw a traditional relay circuit for a simple process, for example starting and stopping a motor or changing its rotation (in clockwise or counterclockwise direction), but if the process is a bit complicated, drawing a direct relay circuit or contactor circuit will be very difficult. By reading this book you can easily draw PLC relay logic for very complex processes. Two advanced PLC programming methods, called the FSM Diagram Method and the Petri Net Method, are discussed with several practical examples. It also provides a new perspective on PLC programming. The presentation process is as follows:

Chapter 1 is about the introductory introduction to the PLC.

In Chapter 2, an introduction to sequential control and the various elements of RLL relay logic are discussed and several practical examples of the RLL program are presented.

In Chapter 3, the structured sequential control design (FSM diagram method) is presented and it is recommended to study this chapter very carefully.

In Chapter 4, the application of Petri nets in PLC programming is discussed, and the design of Petri nets for automation processes using PetriLLD software is described and it is recommended that this chapter be carefully studied.

In chapter 5, hardware components of PLCs that are common to almost all PLCs and several types of industrial sensors are discussed.

Chapter 6 discusses how to work with analog and HMI signals using SIMATIC S7 TIA Portal software.

Chapter 7 discusses how to implement PID control and fuzzy control in PLC environment.

Chapter 8 examines the applications of Ldmicro and PetriLLD in microcontroller programming.

In chapter 9, low cost Arduino based PLCs are discussed. Ladder logic and sequential function chart programming methods for Arduino boards are described.

In chapter 10, Modbus RTU networking and low-cost HMI implementation using touch screen TFT LCDs with Arduino boards are studied.

Finally, in chapter 11, there are several issues of industrial control problems, and the reader is expected to be able to solve all of these problems by using FSM diagram or Petri net methods after reading this book.

I hope this book is a good reference for learning PLC programming. Attempts have been made to provide different examples of PLC programming methods. It is certainly not possible to present a complete list of PLC applications in a book and it is only you can discover the real applications of PLC programming with the help of this book.

Chapter 1

Introduction to PLC

1.1 Introduction

The term PLC means programmable logic controller. The PLC is a software controller that receives data in binary or analog input and processes it in a program that is stored in its memory, and outputs the operation from the output to command executions by sending the command. In other words, PLC is a logic controller that can be defined by the program logic and can be easily changed if needed. The task of the PLC was previously handled by the relay circuits and contactors that have become obsolete today. One of the major drawbacks of these relay circuits was the vast amount of relays were causing the increased volume and weight of these circuits and consequently their prices were increasing and eventually the troubleshooting of these circuits becoming very complex and time consuming. Electronic circuits were developed to address this problem, which were less efficient due to their single-use and need for major modifications to several circuits. Using the PLC changes the process of production or operation of the machine easily because it no longer has to change the wiring and hardware of the control system, and just write a few lines of program and send it to the PLC to accomplish the desired control. PLCs as shown in Figure 1.1 are now widely used in industrial process control technology. The PLC is an industrial computer capable of programmable control operations. Other advantages of PLCs include easy programming and installation, high speed control, network compatibility, troubleshooting and testing, and high flexibility. PLCs are designed for multiple inputs and outputs, wide temperature range, safe from electrical noise and resistant to vibration and mechanical impacts. The control and operation programs of machinery and industrial processes are stored in non-volatile memory with backup battery. The PLC is a real-time system because the output of the PLC-controlled system depends on input conditions, so the PLC is essentially a digital computer designed for use in machine control. Unlike PCs, PLCs are designed for operation in the industrial environment and have specific input/output interfaces and a control programming language. PLC was initially used instead of relay logic but has recently been used in very sophisticated applications. Since the PLC structure is based on the same principles of computer architecture, it not only is capable of performing relay switching but can also perform other operations such as scheduling, counting, computation, comparison and processing of analog signals. Relays require hardware wiring to perform specific operations. With changing the system, relay wiring must be changed and improved.

(a) (b)

Figure 1.1 Programmable logic controllers.

In many cases, such as the automotive industry, the entire control panels have to be replaced because it is not economically feasible to retrofit the old panels. The PLC has eliminated the hardware relation in relay control circuits (Figure 1.2), which is comparatively small and inexpensive compared to relay circuits process control systems. Modern control systems also use relays, but rarely for logic operations. In addition to cost savings, PLC has other benefits such as high reliability, greater flexibility, telecommunication capability, faster response time and easier troubleshooting.

1.2 History of PLC

The idea of making PLC was first introduced in 1968 by a group of engineers from General Motors Corporation of US. In this design, the controller should have the following basic characteristics:

1. It is easily programmable as well as reprogrammed, prefera-bly in the factory, has the ability to change the order and se-quence of control operations.
2. Easy to maintain, preferably using add-on modules.
   a) More reliable in industrial environments
   b) Smaller than its equivalent relay circuit

3. Practically competitive with semiconductor relay circuit boards.

This has sparked widespread enthusiasm among engineers of all sciences about how to use PLC in industrial controls. This attention was drawn to the PLCs superior capability and facilities, which made it rapidly becoming available technology. The instructions also evolved rapidly from simple logic commands to instructions including counters, timers, register shifts and then advanced mathematical functions in larger PLCs.

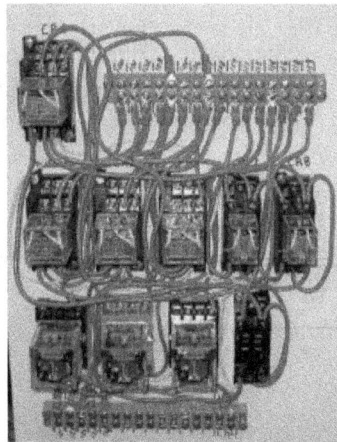

(a)

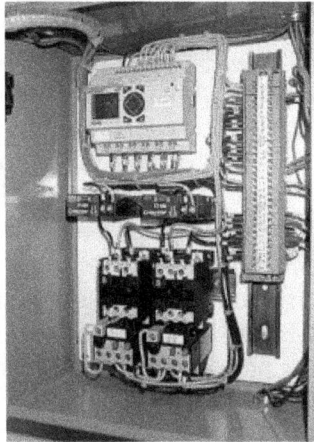

(b)

Figure 1.2 Relay and PLC control panels (a) relay control panel and (b) PLC Control panel.

In parallel, improvements to the PLC hardware followed with larger memory and more inputs and outputs on the newer modules. In 1976 it was possible to control remote input/output modules. In such applications, many of these inputs/outputs, which were several hundred meters away from the PLC, had to be continuously monitored through a single line of communication or the necessary instructions were given. In 1977, a microprocessor based PLC was introduced by the microprocessor-based PLC (American company Alan Bradley), an 8080 microprocessor based PLC, However, used another processor to handle high bit rate logic instructions. The application of PLCs in industries encourages producers to expand and develop different microprocessor-based systems with different levels of operation, and today has the range of available PLCs from comprehensive small complete PLCs with 20 input/output and 500 programming steps to modular systems with expandable modules. Modules are used to perform tasks such as the following:

- o   Analog input/output
- o   Triangular PID control: proportional, integral, and derivative
- o   Communications
- o   Graphic representation
- o   Extra input/output
- o   Extra memory

PLC modularization solutions allow for the development or improvement of a control system at minimal cost and drawbacks. Today, PLCs are almost at the same pace as microcomputers, with the emphasis being on controlling such as numerical/positional control and communication network capabilities. From a market standpoint, the market for small controllers has seen rapid growth since the early 1980s as a number of Japanese companies introduced low-cost PLCs from other products at the time. They were cheaper because potential customers in the industry were able to buy programmable controllers. This trend continued with the supply of more efficient PLCs as cheaply as possible. Table 1.1 illustrates the emergence and development of PLCs over the past years.

Table 1.1 The emergence and development of PLCs over the past years

Year	Aspects of Progress
1968	Introducing the idea and concept of programmable controllers
1969	Hardware CPU controller with logic commands and 1 kB of memory and 128 input/output terminals
1974	Using multiple processors with one, PLC timers and counters, computational operations, 12 kB of memory and 1024 input/ output terminals
1976	Introducing Long Distance Input/Output Systems
1977	Introduction of microprocessor based PLCs
1980	Development of intelligent input/output modules Improvement of communication facilities Improved software properties e.g. document ability
1983	Introducing small and low cost PLCs
1985 onwards	Distribution of all surfaces, PLC computers and machines under GM standard and distributed control and hierarchy of industrial plants.

1.3 Content presentation process

The first step in automation of industrial processes and machinery is to determine the sequence of events specific to their performance. This order is arranged in a set of logical functions. Logical functions are of two types:

a) Combined: Results are only dependent on the current state of the inputs.
b) Sequential: The results are dependent on the current and previous state of the inputs.

This logical scheme is transformed into a physical system using specific structural and technological blocks, namely mechanical, hydraulic, pneumatic, electromechanical, and electronics. In combined control systems, the inputs send information to the system at any given moment and the system outputs are controlled at any given moment. A change in input directly causes a change in output. Conversely, in a sequential control system, a series of different events occurs one after the other. Completing an event in a sequence provides a signal for the next event to begin. Examples of sequential systems are:

    a.   Timers for controlling central heating systems
    b.   Laundry machines
    c.   Traffic light
    d.   Building elevators, etc.

Sometimes one of the events in the sequential control itself is a combined control system. For example, filling the washing machine with water uses a combined control system that monitors the water level and controls the inlet valves. However, this is just one event in a series of events that provides the complete sequential control system for the washing machine. Sequential control using ladder logic, PLC and operator interfaces are widely used in manufacturing industries. PLCs have been programmed since the 1970s using ladder logic. This programming technique is often an empirical process that begins by inserting the first rung to satisfy the first required output. The rungs continued until the problem was resolved. At some point in the programming process, the system does not respond to the desired test response time, and rungs need to be corrected to resolve the problem. This process continues until the program is resolved. As a result, it is an unstructured sequential program that is difficult to analyze and understand. In large PLC applications, it is very difficult to troubleshoot the program when system failures occur. A technique called Sequential Function Chart (SFC) is used in sequential PLC programming. The SFC process adds a structure to the PLC programming process that makes it easy to analyze and troubleshoot.

PLC programming uses two techniques: experimental and structured. The experimental method includes:

1.   A clear statement and definition of the control problem
2.   Determine all control outputs and control needs for each output
3.   Determine all combinations of the input conditions for each control output needed to generate an active output state
4.   Create ladder rungs for each control output using the input conditions specified in step 3
5.   Program scans from first rung to last rung until outputs produce optimal system performance with input modes switching
6.   Adding rungs to correct step 5 problems

    7.   Documentation of ladder logic by identifying all inputs and outputs and labeling the program and all its rungs

The experimental method described in the above seven-step process works well in applications of small ladder logic. However, as the control problem becomes more complex, the number of problems in step 5 increases and the number of steps to resolve the problem increases. Therefore, it is very difficult to read and understand the program. Another problem with this method is the inconsequential nature of program structure. When PLC scans the program from the first rung to the last rung, it triggers the outputs based on the input logic. The sequence change in output states is not necessarily followed. For example, the final rung output might first triggered and looked for another rung output. A large control system may have a ladder with several thousand rungs. It is very difficult to determine the order of output and machine performance of such a large system if using an experimental technique. Structured PLC programming provides a mechanism by adding structure to ladder logic programming. The timing to do and the number of rungs are increased but the execution time is reduced. In addition, it is easier to follow the program and troubleshoot the program when the problem arises, which is discussed in chapters 3 and 4 with FSM (Finite State Machine) and Petri nets programming methods. The presentation process is as follows:

Chapter 1 is about the introductory introduction to the PLC.

In Chapter 2, an introduction to sequential control and the various elements of RLL relay logic is discussed and several practical examples of the RLL program are presented.

In Chapter 3, the structured sequential control design (FSM diagram method) is presented and it is recommended to study this chapter very carefully.

In Chapter 4, the application of Petri nets in PLC programming is discussed, and the design of Petri nets for automation processes using PetriLLD software is described and it is recommended that this chapter be carefully studied.

In chapter 5, hardware components of PLCs that are common to almost all PLCs and different types of industrial sensors are discussed.

Chapter 6 discusses how to work with analog and HMI signals using SIMATIC S7 TIA Portal software.

Chapter 7 discusses how to implement PID control and fuzzy control in PLC environment.

Chapter 8 examines the applications of Ldmicro and PetriLLD in microcontroller programming.

In chapter 9, low cost Arduino based PLCs are discussed. Ladder logic and sequential function chart programming methods for Arduino boards are described.

In chapter 10, Modbus RTU networking and low cost HMI implementation using touch screen TFT LCDs with Arduino boards are studied. Finally, in chapter 11, there are several issues of industrial control problems, and the reader is expected to be able to solve all of these problems by using FSM diagram or Petri nets methods after reading this book.

Chapter 2

Introduction to Sequential Control

2.1 Introduction

Sequential control is a class of control issues for systems in which inputs, outputs and feedbacks are discrete (on/off) values.

- Specific output values
- Certain amounts of output that are subject to time constraints
- The order of the outputs
- Sequence between different outputs

For example, in Figure 2.1, the drill starts at the highest point where the LS1 limit switch is stimulated. When button is pressed the cycle starts, the power on lamp illuminates and the clamp is stimulated and the drill goes down (with Head-Z stimulation) and the spindle motor is triggered. When reaching the LS2 limit switch, Head-Z is turned off and Head+Z is triggered. Finally, when reaches the LS1 limit switch, the clamp is released and the spindle motor and Head+Z are switched off. The system is reset by pressing the stop button at any time.

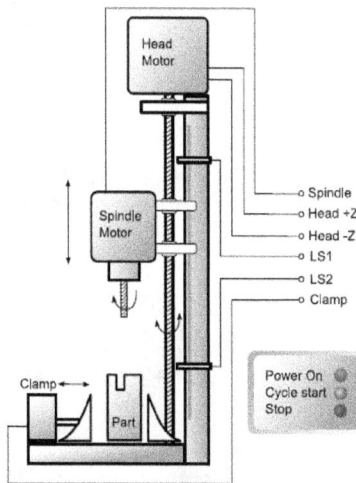

Figure 2.1 An example of sequential control.

Two types of modeling for systems are CV (Continuous Variable) and DE (Discrete Event) models.

1. CV model:
- Real values and unlimited state space.
- Continuous or discrete time is used. Differential equations are used to study continuous time systems while difference equations are used in discrete time systems.

2. DE Model:
- Discrete values and finite state space.
- Time is synchronous or asynchronous. In synchronous mode, the values change with clock while in asynchronous mode the values change at any time.

2.2 DE modeling method

- Variables: input, output and state
- The number of state variables is limited
- State change occurs for the following reasons:
    a) External events (inputs)
    b) Internal dynamics
- Events (outputs) occur due to a change of states
- The system specifications are as follows:
    a) The activation conditions are required for the transition
    b) The sequences are transitions/events
    c) Transitions are temporal/non-temporal

For example, we have in Figure 2.1:
- Inputs: Cycle Start, Stop, LS1, LS2
- Outputs: Power on, Spindle motor, Clamp, Head-Z, Head+Z
- States: state1 (Idle), Power on, Spindle motor on, Clamp on, Head-Z on, Head+Z on

State changes occur for the following reasons:
- External Inputs: Cycle Start, Stop, LS1, LS2
- Internal Dynamics: move up Head+Z to LS1 and move down Head-Z to LS2

Events (outputs) are caused by state changes. Figure 2.2 shows a diagram of its states.

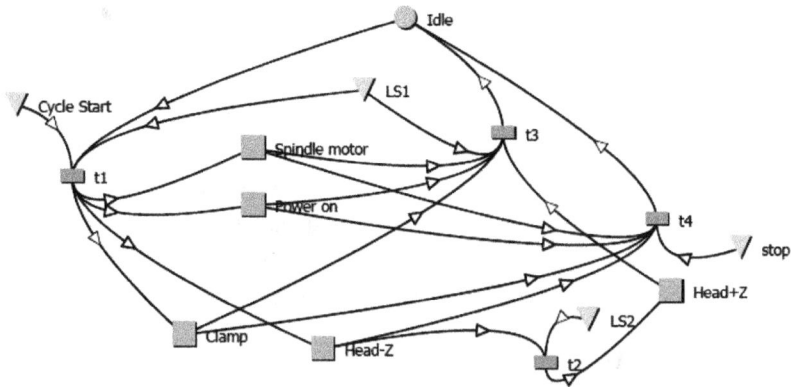

Figure 2.2 System states diagram of Figure 2.1.

2.3 Comparison of sequential control and analog control

Items	Sequential Control	Analog Control
Process Variables	Discrete value	Continuous value
Model	Logical and state-transition	Numerical and differential equations
Signal	Sequence/state	Signal value/time function
Open/Closed Loop Control	Logical on/off and supervisory	Linear/nonlinear and automatic
Design	Simple	Complex
Settings	No need due to slight changes in the process model	It is needed and the process model is prone to change

2.4 Comparison of PLC and Relay Logic

Relay Logic	PLC
Hardware logic using relays and switches	Software logic using CPU and memory
It is difficult to update and repair it	It is modular and easy to program
It has speed, size, complexity and reliability limits	It's great in these cases
It's an old technology	It's an up to date technology

2.5 Definition of PLC (Programmable Logic Controller)

PLCs are a type of logical controller from the family of computers designed and built for industrial applications. PLCs are used to automate operations in the manufacturing production plant. The main task of a PLC is to get information from the unit under control as input to the system, to decide on the input values and to program in which it is embedded, and ultimately to generate outputs and send them to intermediate hardware to drive controlled machines.

2.6 Components of a PLC

All PLCs have a variety of hardware to perform control operations, each of which performs a specific task. The following elements are present in all PLCs.

- Back-Plane, Power Supply (PS), CPU, Memory, I/O Cards
- Digital and analog modules
- Functional Modules
- High speed counters
- Positioning Modules
- wiring
- Add-on components
a) Programmer
b) MMI (Man Machine Interface)

2.7 Specifications of PLCs

- Real-time system
- Embedded control
- Sequential control hardware
- Specific programming methods
a) Graphic
b) RLL (Relay Ladder Logic)
c) Block diagram etc.
The most important PLC programming method is RLL, which is used as the main programming language in all PLCs.

2.8 RLL (Relay Ladder Logic)

The components and methods of the RLL programming are illustrated in Figure 2.3.

Simple RLL elements

1. Input contacts: External and real.
a) NO (—| |—) contacts: When powered up, is closed and opened normally.
b) NC (—|/|—) contacts: When powered up, is opened and closed normally.
2. Auxiliary contacts: They operate when the output coil is energized.
3. Output Coil: Indicates the operation of the output device (lamp, motor, etc.).

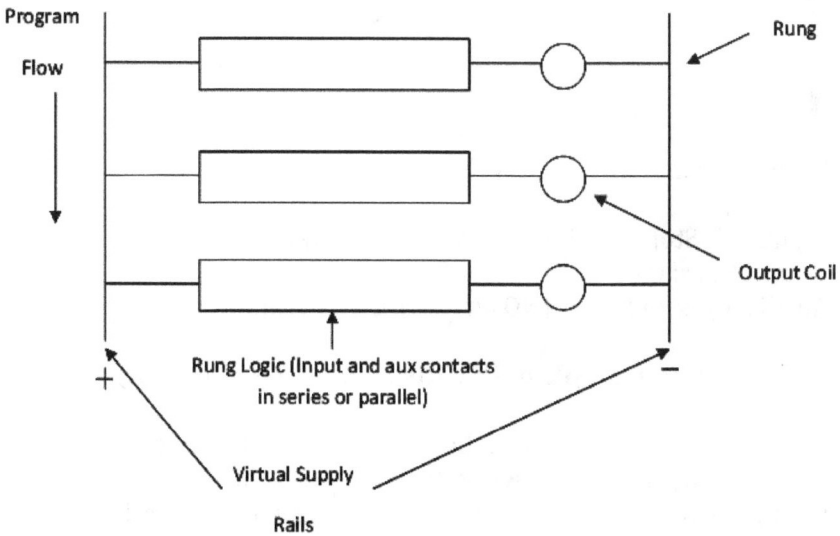

Figure 2.3 RLL programming components and methods.

2.9 Example of RLL (Left-Right Rotation Control of a Motor)

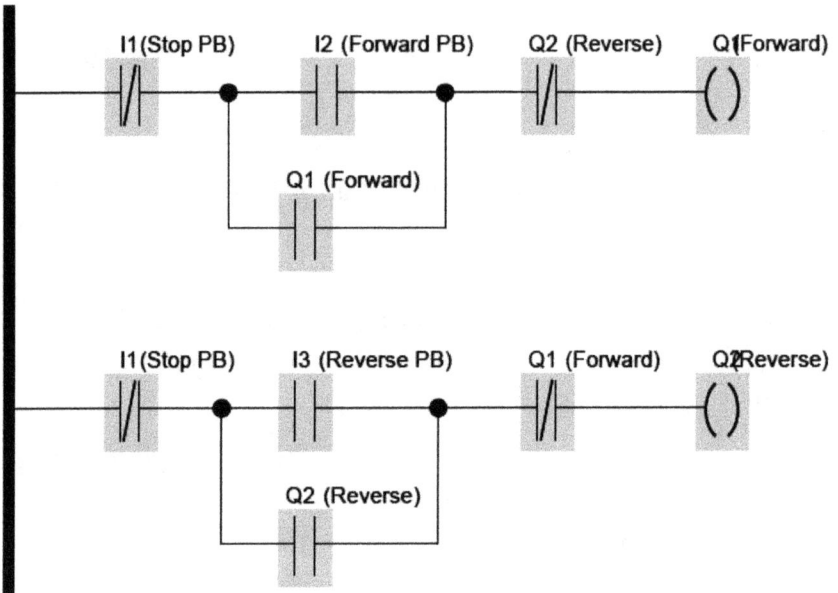

Example elements:

Inputs: I1: Stop PB, I2: Forward PB, I3: Reverse PB
Output Coils: Q1: Forward, Q2: Reverse
Auxiliary Contacts: Q1 (NO and NC), Q2 (NO and NC)

2.10 Another example of RLL (hydraulic press control)

Figure 2.4 shows the hydraulic press machine under control. First, suppose the press is in down state. In this case, the Dn-LS limit switch is stimulated then Up-Sol is triggered and the press is raised to the Up-LS limit switch and Up-LS is stimulated then Dn-Sol is triggered. The press is lowered to the Dn-LS limit switch and Dn-LS is stimulated, so Up-Sol is triggered and the press goes up to the Up-LS limit switch, and this cycle continues. The system's RLL program is as follows.

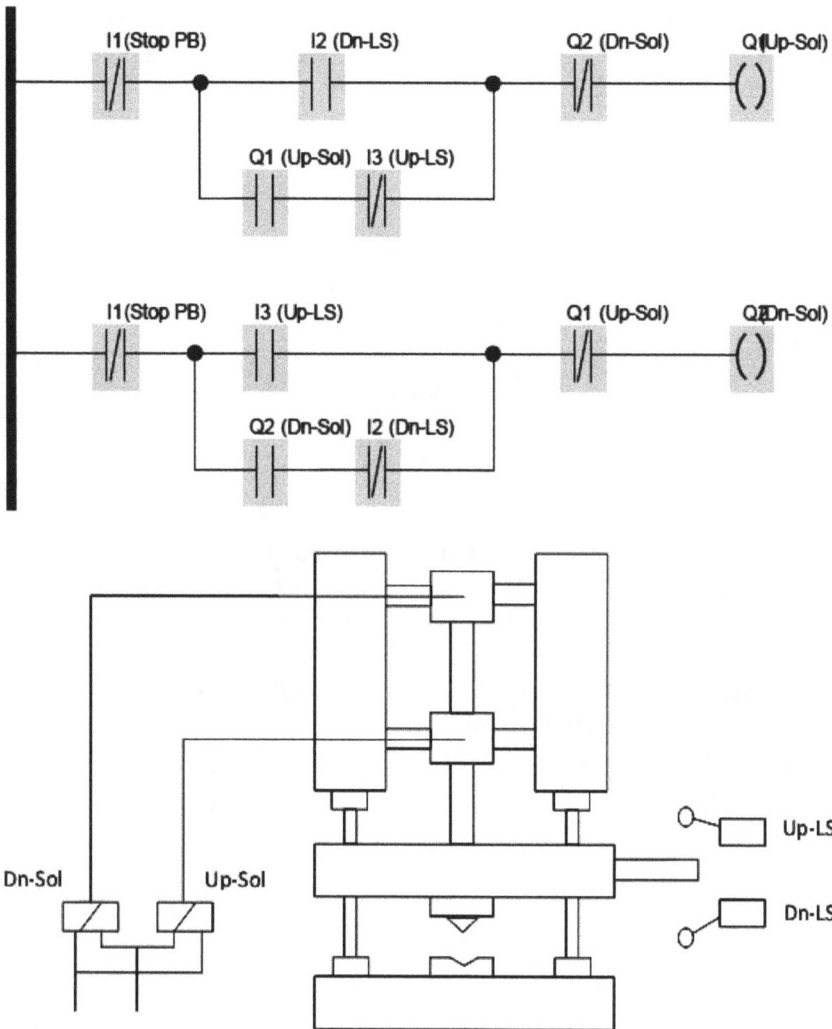

Figure 2.4 Hydraulic press machine.

2.11 Other RLL elements

Timers:

Timers are created to calculate the output time of the timers, based on their type and according to the time specified for them. Thus, a start base activates its function, calculates the desired time and produces a proportional output depending on the timer type. The overall structure of the timers is shown in Figure 2.5.

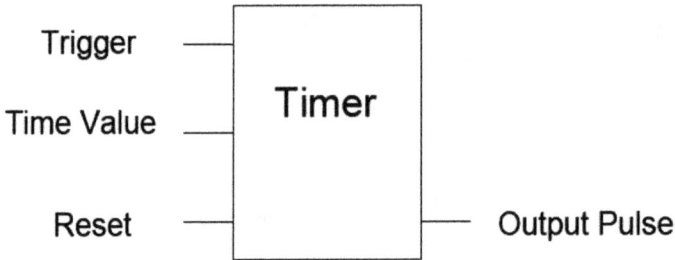

Figure 2.5 The overall structure of the timers.

By applying a "Positive Edge Trigger" or shifting from "0" to "1" on the trigger base, the timer is activated and started, and a specific pulse appears at its output. The shape and size of the pulse depends on the timer type. It is also given a measurable amount of time through the Time Value base. A reset base is also used to stop the timer. By applying a rising edge on this base, the timer is inactive and its output is zero. A variety of timers may be used for programming in PLCs. The most common types of timers that are available and used in most PLCs are:

- On-delay Timer
- Off-delay Timer
- Pulse Timer
- Fixed Pulse Timer

On-delay Timer:

The timing diagram for this timer is shown below:

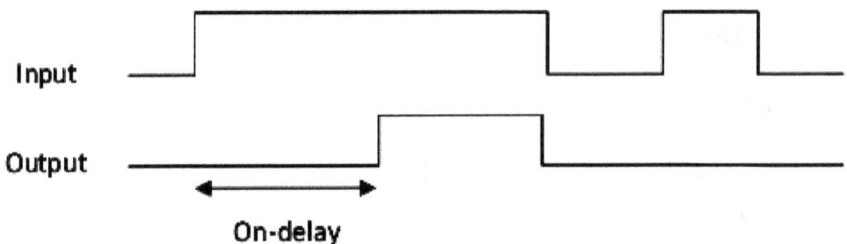

Off-delay Timer:

The timing diagram for this timer is shown as follows:

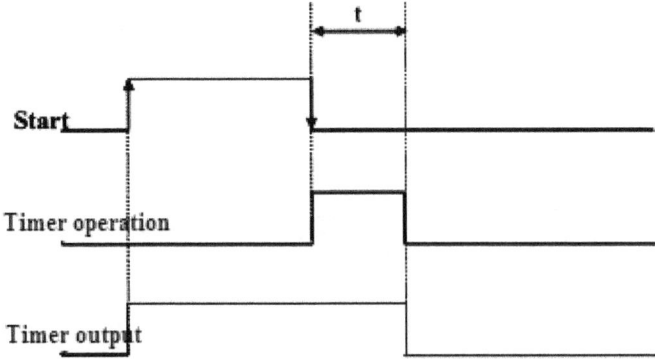

Pulse Timer:

The timing diagram for this timer is shown below:

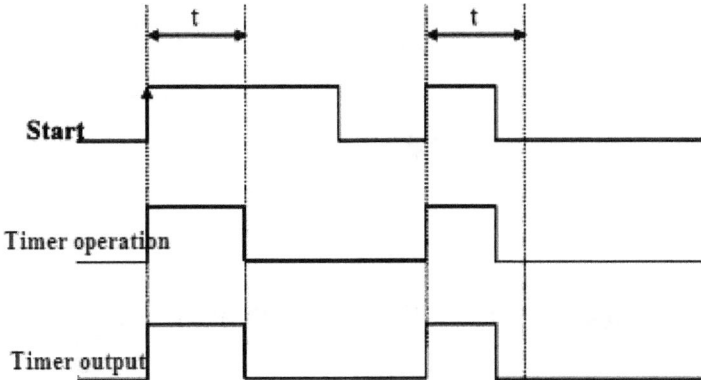

Fixed Pulse Timer:

The timing diagram for this timer is shown as follows:

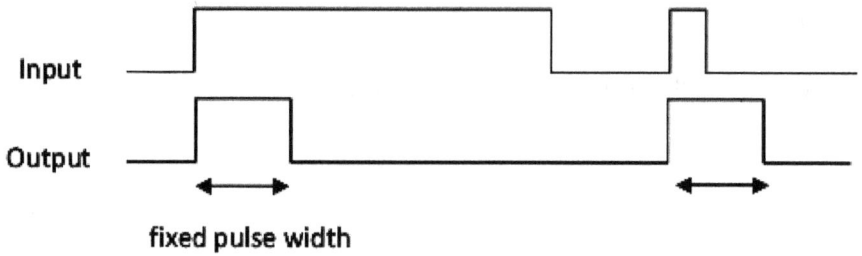

fixed pulse width

Timers in other words:

a) Non-retentive
b) Retentive

Non-retentive Timer:

The timing diagram for this timer is shown below:

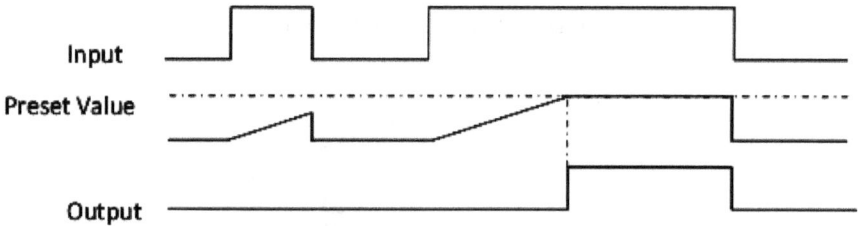

Retentive Timer:

The timing diagram for this timer is shown as follows:

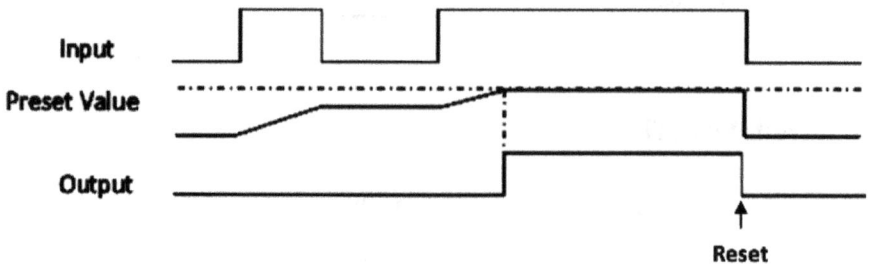

2.12 An example of timers

If in the example of a hydraulic press you want the press be in down position for 5 seconds and then go to the up position then RLL program is written as follows.

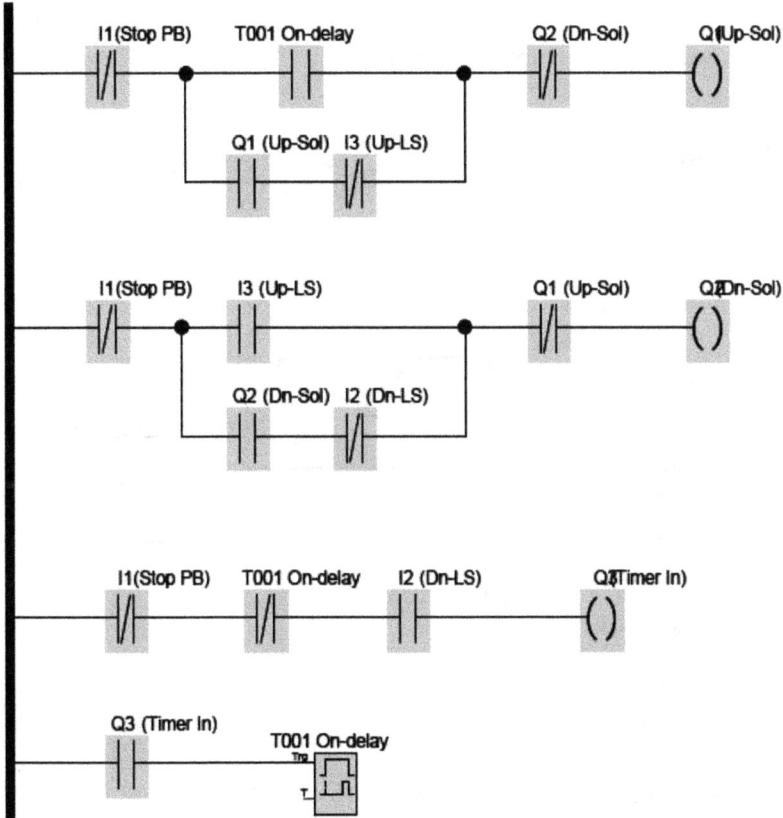

2.13 Counters

The action of counters is to count the number of input pulses. Counters usually have one input to determine the number of counts and another input to activate the counter. The overall structure of the counters is as shown in Figure 2.6. Counters are widely used in PLC programming. In most PLCs there are two types of counters (up-counter and down-counter).

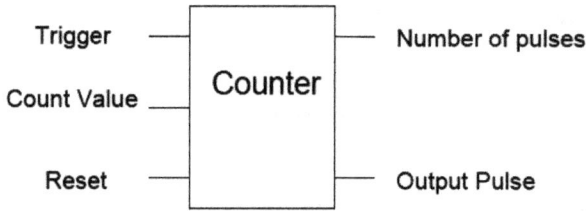

Figure 2.6 Overall structure of the counters.

Up Counter:

The timing diagram of this counter is shown below:

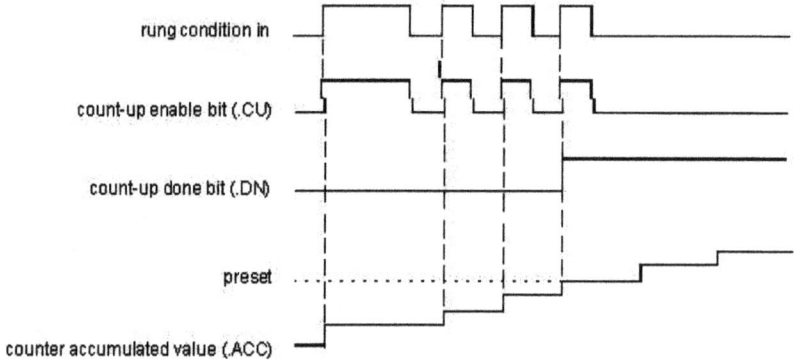

Down Counter:

The timing diagram of this counter is as follows:

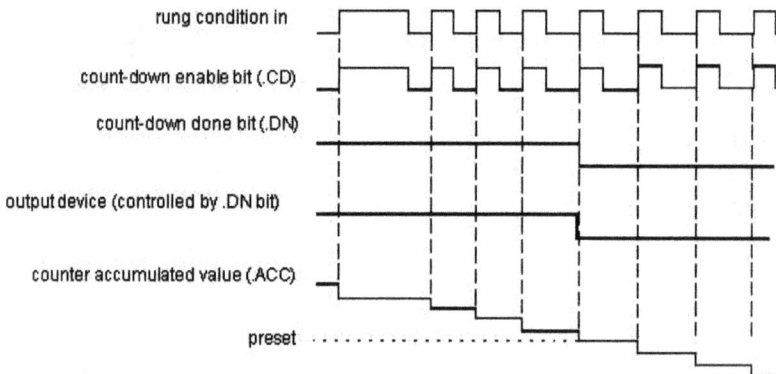

Up/Down Counter:

The timing diagram of this counter is shown below:

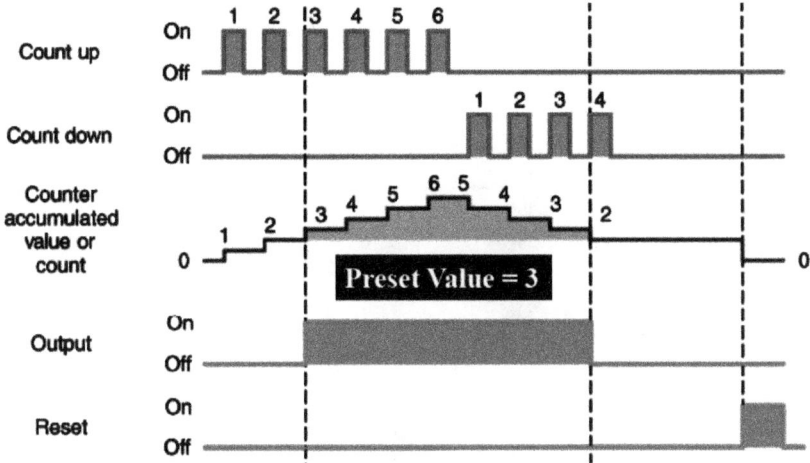

2.14 An example of counters

The conveyor motor M operates when x numbers of box A and y numbers of box B be on conveyor. This system is illustrated in Figure 2.7.

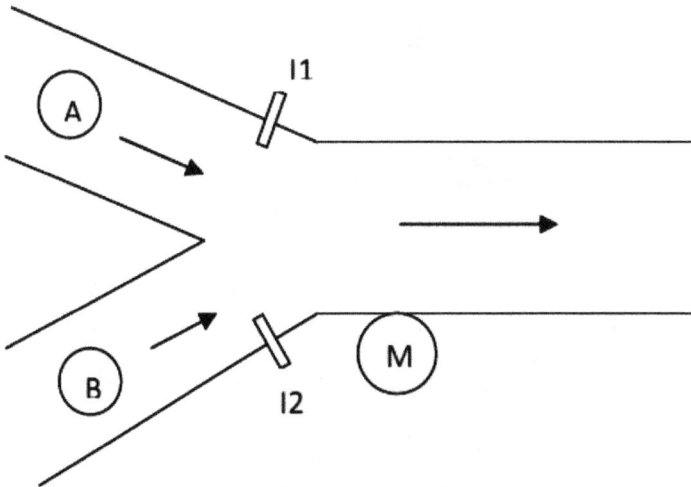

Figure 2.7 Controlling the conveyor motor using counters.

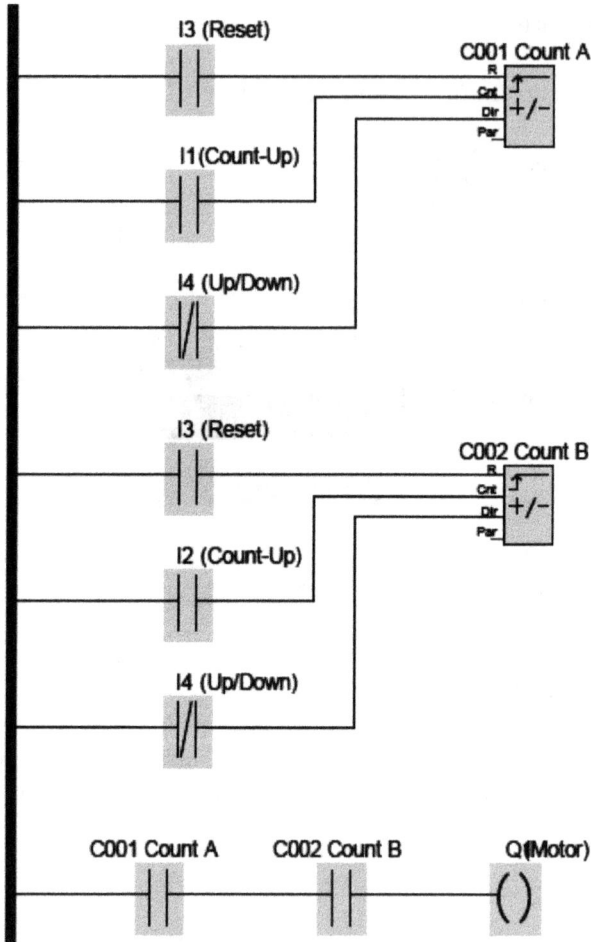

2.15 Another example of counters

By pressing the start button, the conveyor motor starts. The boxes move over the conveyor, passing the proximity switch, increasing the counter count. After counting the boxes up to 50, the conveyor motor automatically stops and the counter count resets to zero.

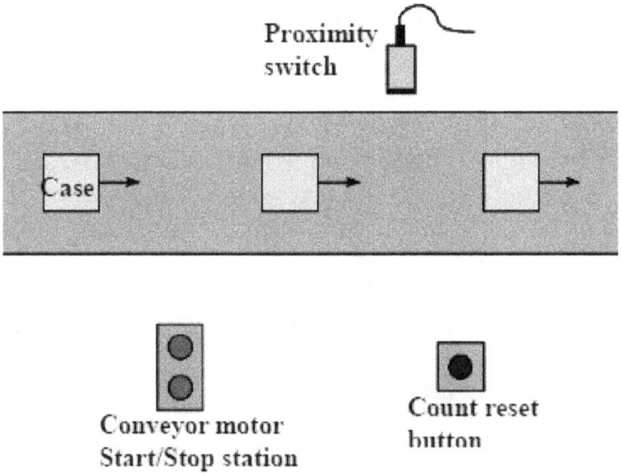

Figure 2.8 Conveyor motor control system using counter.

The conveyor motor can be manually start or stop operated at any time without losing count. This system is illustrated in Figure 2.8.

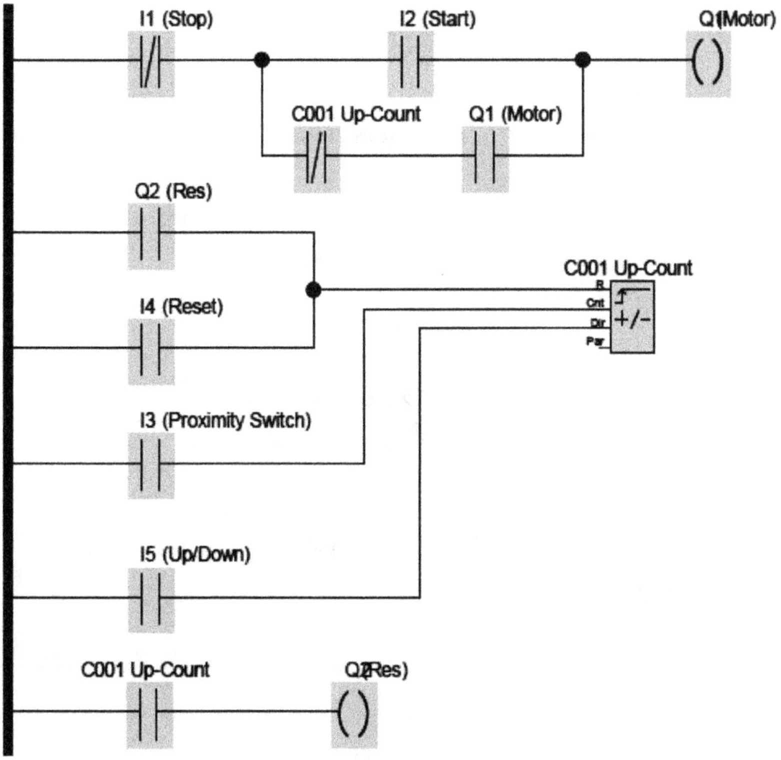

2.16 Another example of counters

When a car enters the parking lot, a number is added to the Up-Counter value, and when the car exits, the number is reduced from a Down-Counter value. When the counter is reaching to 150, the counter output is activated and the Lot-Full lamp lights up.

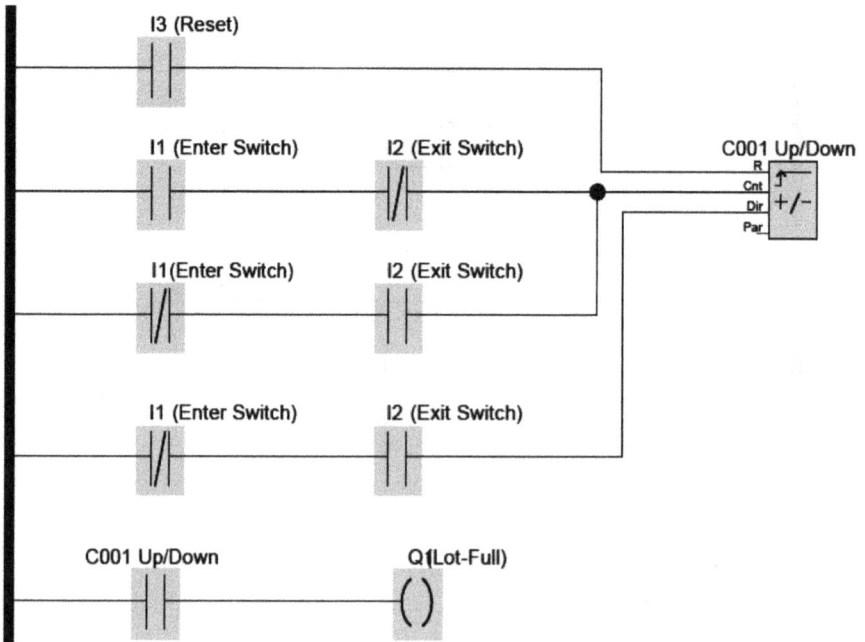

2.17 Example (Counter and Timer Combination)

When the start button is pressed, the M1 conveyor starts to move. After 15 plates overlap on conveyor M2, the M1 conveyor stops and the M2 starts moving. After the M2 conveyor has been operated for 5 seconds, the process is repeated automatically. The timer output resets timer and counter and provides an instant pulse to restart the M1 conveyor. By pressing the stop push button, both the M1 and M2 conveyors stop. This system is illustrated in Figure 2.9.

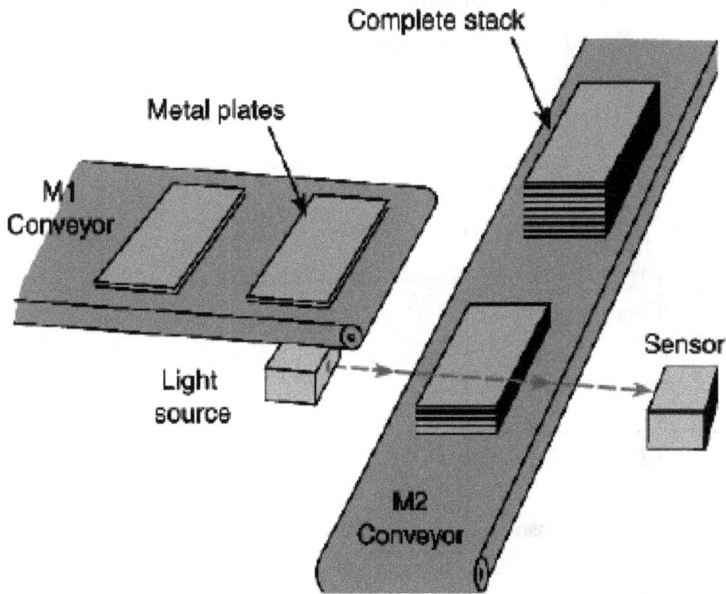

Figure 2.9 Example of Counter and Timer combinations.

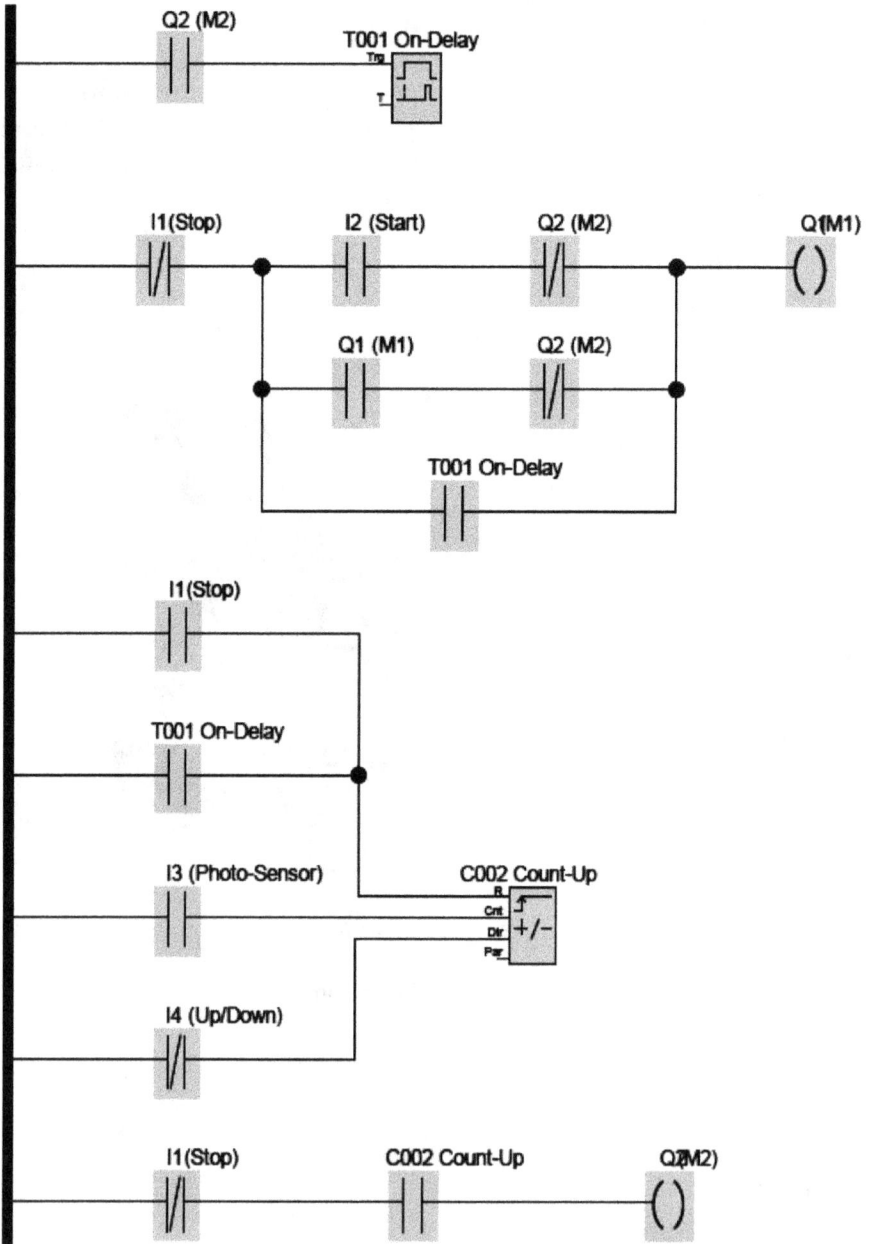

2.18 Example (Counter and Timer Combination - Production Rate)

When the start switch is closed, both timers and counters are activated. The counter counts for each piece that passes in front of the sensor. The counting operation is performed and the timer operates for 1 minute. The value counted in a minute by the counter indicates the production rate. This system is shown in Figure 2.10.

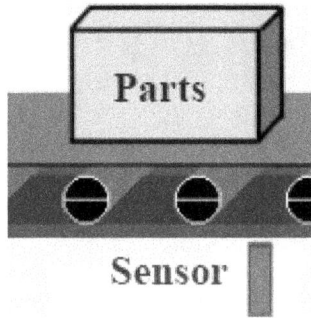

Figure 2.10 Another example of Counter and Timer combinations.

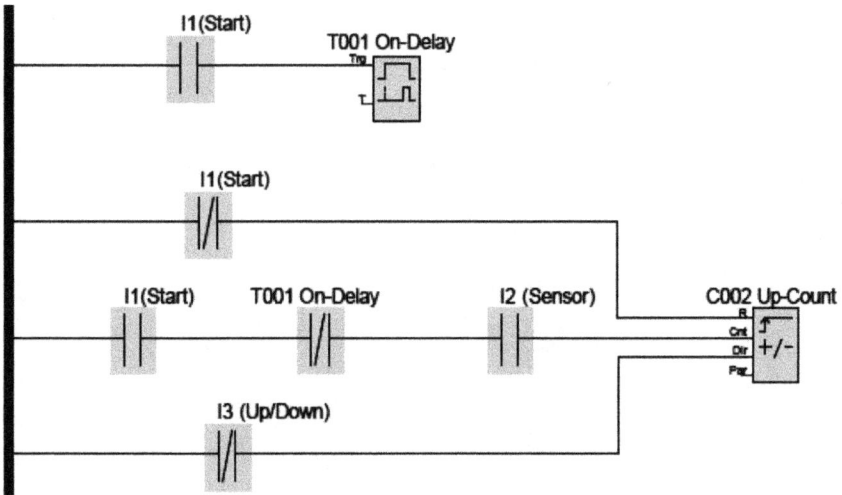

Chapter 3

Structural Sequential Control Design

3.1 Introduction

Structured design is very useful for finding the fault and also designing a fault-free control system. In this topic, sequential control is modeled using state machines and the process of converting state machines into RLL applications is explained. The basic structures of a SFC (Sequential Function Chart) program are also described.

3.2 Designing steps for sequential control

A) Study the behavior of the system

1) Determination of inputs from sensors and MMI (Man Machine Interface)
2) Determination of outputs for operators and MMI
3) Study the sequence of operations and events under different functional modes
4) Study the effects of possible failures
5) Examine the need for manual control, additional sensors, displays, maintenance alarms, performance or safety efficiency.
Note: Manual control is located near the equipment or, in other words, in the field while the PLC control is in the control room.

B) Converting linguistic description into formal process modeling

1) Intermediate forms such as operations list, flowchart, etc. may be used first.
2) Finally, converting the linguistic description into a formal framework such as the FSM (Finite State Machine) model.

C) Design of sequential control logic based on formal model

D) Perform control logic in RLL program form

3.3 Example (Hydraulic Press Process)

The hydraulic press system is illustrated in Figure 3.1.

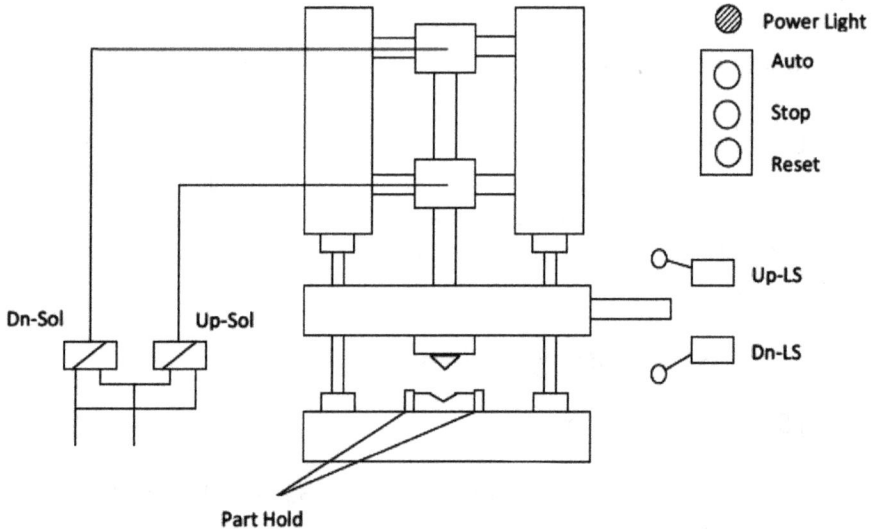

Figure 3.1 Hydraulic press system.

Linguistic description of the hydraulic press process:

A) By pressing the "Auto" PB push button, the system starts and the Power Light lamp lights up.

B) When a part is detected by the Part Detect sensor, the press comes down to Dn-LS.

C) The press then returns to Up-LS and stops.

D) PB "Stop" stops the process only when the press comes down.

E) If the "Stop" PB push button is pressed, the "Reset" PB push button must be stimulated before triggering the "Auto" PB push button.

F) After returning the press, the press waits until the part is removed and the next part is detected by the Part Detect Sensor.

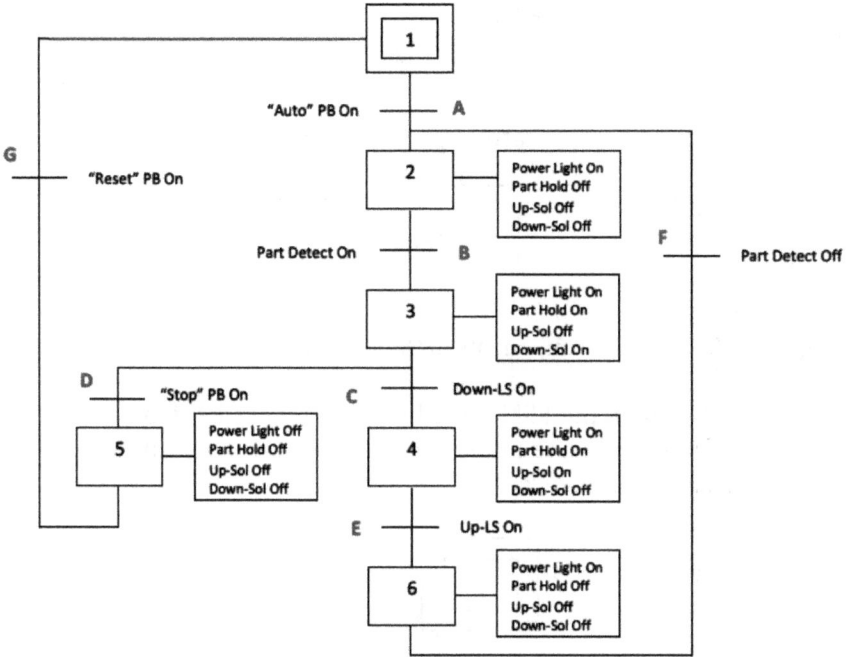

Figure 3.2 FSM model.

Mathematical Description

Inputs:
- Part Detect
- Auto PB
- Stop PB
- Reset PB
- Up-LS
- Down-LS

Outputs:
- Up-Sol
- Down-Sol
- Part Hold
- Power Light

The FSM (Finite State Machine) model of the hydraulic press process is shown in Figure 3.2. The FSM model consists of three main parts, as shown in Figure 3.3.

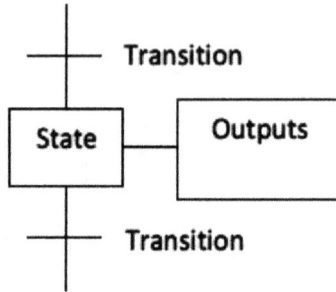

Figure 3.3 The main components of the FSM model.

The initial state in the FSM model is represented by two nested rectangles. The initial state relates to the state from which the system starts, as shown in Figure 3.4.

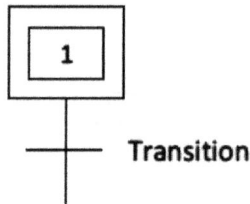

Figure 3.4 Initial state of the FSM model.

The output table for the FSM model is given below.

Outputs	State No					
	1	2	3	4	5	6
Power Light	0	1	1	1	0	1
Part Hold	0	0	1	1	0	0
Up-Sol	0	0	0	1	0	0
Down-Sol	0	0	1	0	0	0

To convert the FSM model to the RLL program, we form the following three RLL diagrams for Transitions, States and Outputs.

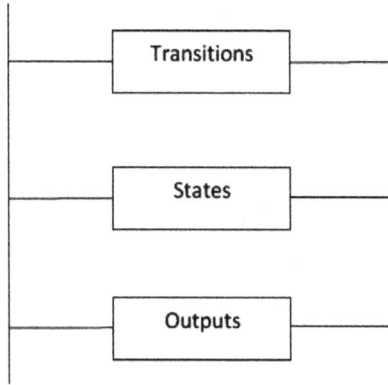

Transition Logic

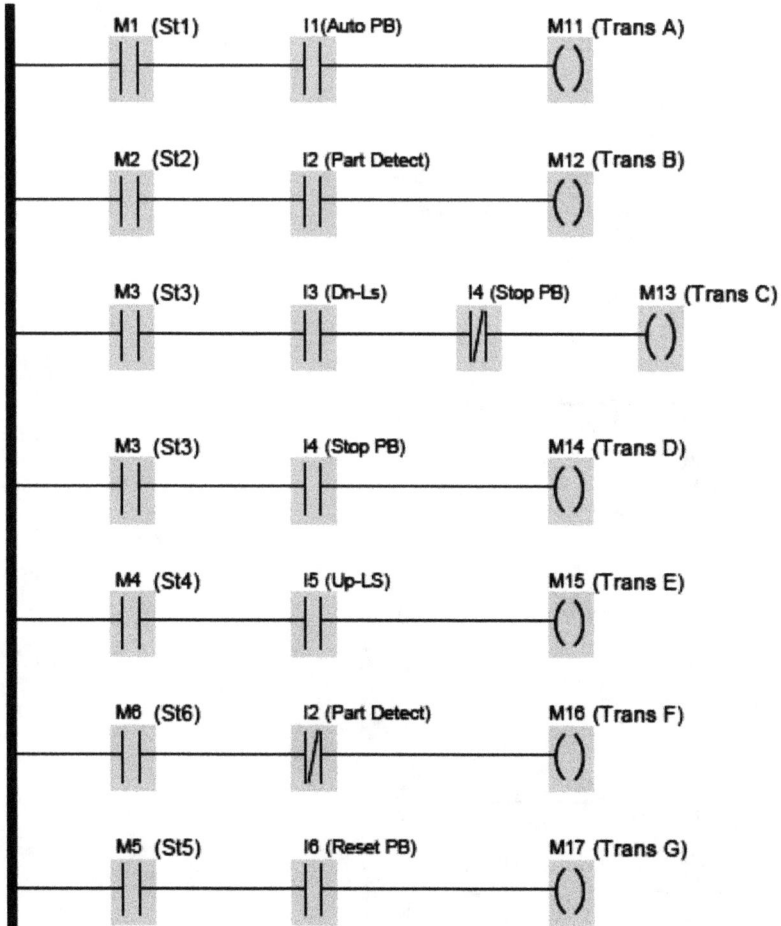

State Logic

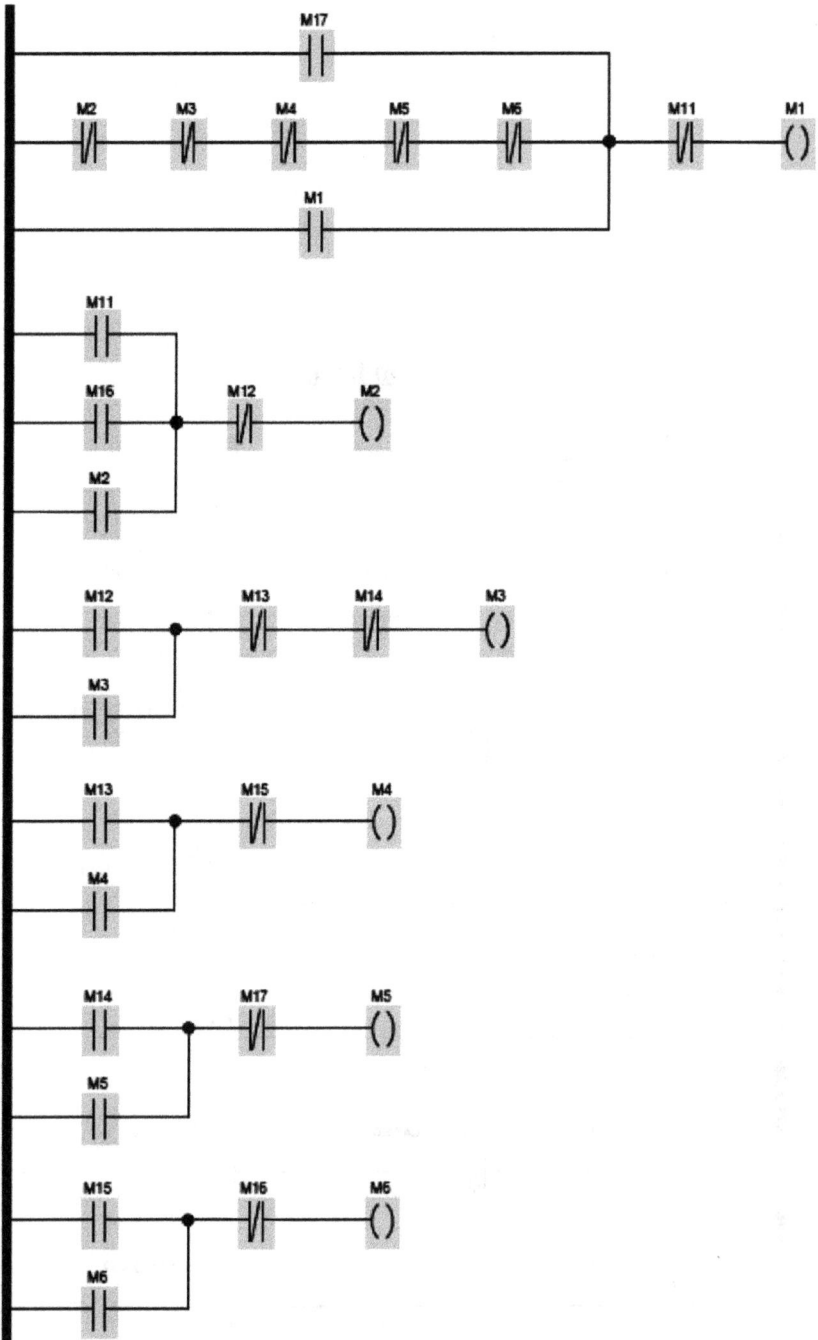

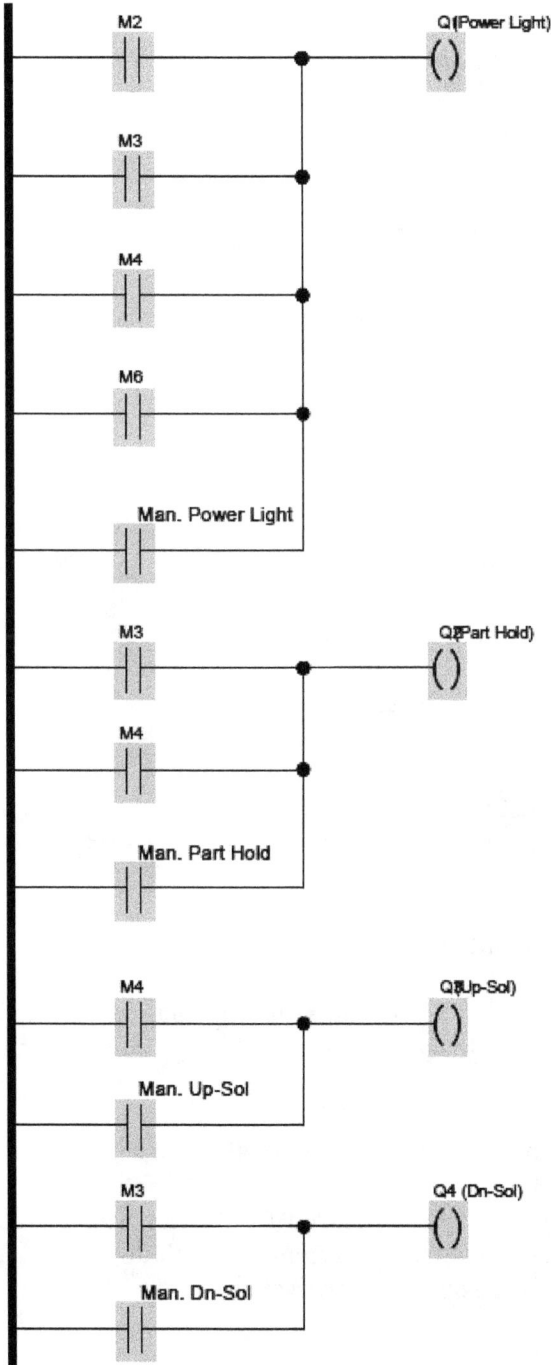

Output Logic

	M2		Q1 (Power Light)
	M3		
	M4		
	M6		
	Man. Power Light		
	M3		Q2 (Part Hold)
	M4		
	Man. Part Hold		
	M4		Q3 (Up-Sol)
	Man. Up-Sol		
	M3		Q4 (Dn-Sol)
	Man. Dn-Sol		

3.4 Unstructured RLL programming of the hydraulic press process

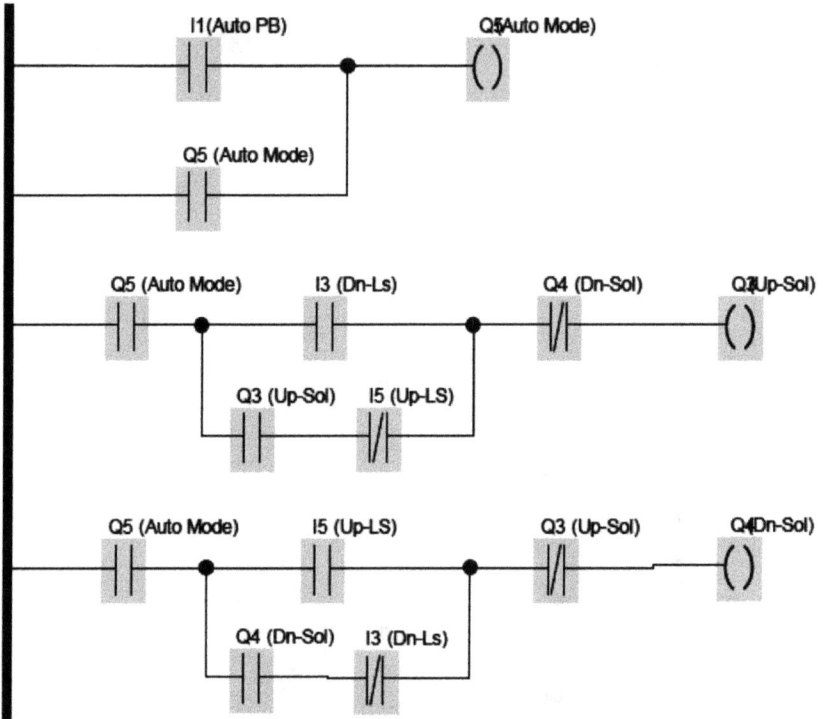

Writing an unstructured RLL program is very difficult and complicated. If for example the control system has 200 states, it is impossible to write this program directly.

3.5 Benefits of advanced programming

According to IEC1131, which is the international standard for PLC programming languages, two types of programming languages have been introduced:

a) Graphic languages: FBD (Functional Block Diagram) and LD (Ladder Diagram)
b) Text languages: ST (Structured Text) and IL (Instruction List)
The benefits of advanced programming include:
- Its standards are open and accessible.

- Modules are structured as a module, making it easy for program development, maintenance and updating.
- In each program scan, states are calculated in Active Modules only.
- Supports concurrency.

3.6 SFC (Sequential Function Chart)

SFC Definition: A graphical sample for describing program structure with modular concurrent control.
Note: SFC only describes the structured organization of program modules and the programs themselves must be written in one of the available programming languages.

Basic SFC Structures

A) States: Each state is a control module that can be programmed as RLL or any other program. There are two types of states: Initial state and Regular state. The initial state is performed on the first run and after the reset, and the normal state occurs when the transition logic is enabled. When a state is disabled, its state is initialized and only active states are evaluated during scanning.

B) Transitions: Each transition is a state-controlled control module that ultimately evaluates a transition variable. When the transition variable is correct, the following states are activated and the preceding states are disabled. Only transitions followed by active states are evaluated. Transition can be a simple variable value.

3.7 Program control structures

A) Selective/Alternative Branch: This structure is shown in Figure 3.5. If T1 and T2 are activated simultaneously the priority will be left (T1) and the next state will be S2. So if more than one transition variable is true, the priority is left to right. Only one branch is active at a time. If S4 is enabled and T3 is correct, S6 is enabled and S4 will be disabled.

B) Simultaneous/Parallel Branch: This structure is illustrated in Figure 3.6 and is the structure used for concurrent control programming. If S1 is active and T1 is correct, S2 and S3 are enabled and S1 is disabled. If S4 and S5 are active and T2 is correct, S6 is activated and S4 and S5 are disabled.

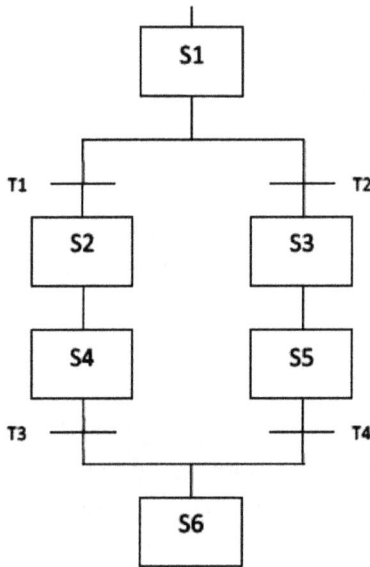

Figure 3.5 Selective/Alternative Branch.

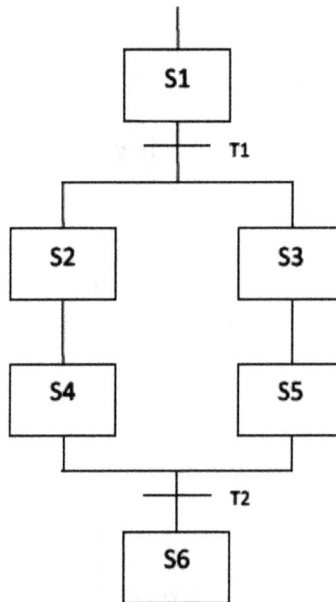

Figure 3.6 Simultaneous/Parallel Branch.

3.8 Example 1

By pressing the start push button pump1 and pump2 are triggered simultaneously when the limit switch1 is activated pump1 is switched off. When the limit switch3 is activated pump2 is switched off. When limit switch1 and limit switch3 are both activated, valve1 and valve2 are opened simultaneously. Valve1 is also closed when limit switch2 is deactivated. When limit switch4 is deactivated valve2 closes. When limit switch2 and limit switch4 are both deactivated the mixer will start for 100 seconds and then valve3 will open. When limit switch5 is deactivated, valve 3 closes and the process is terminated and the start button must be pressed to start the process again. This system is illustrated in Figure 3.7.

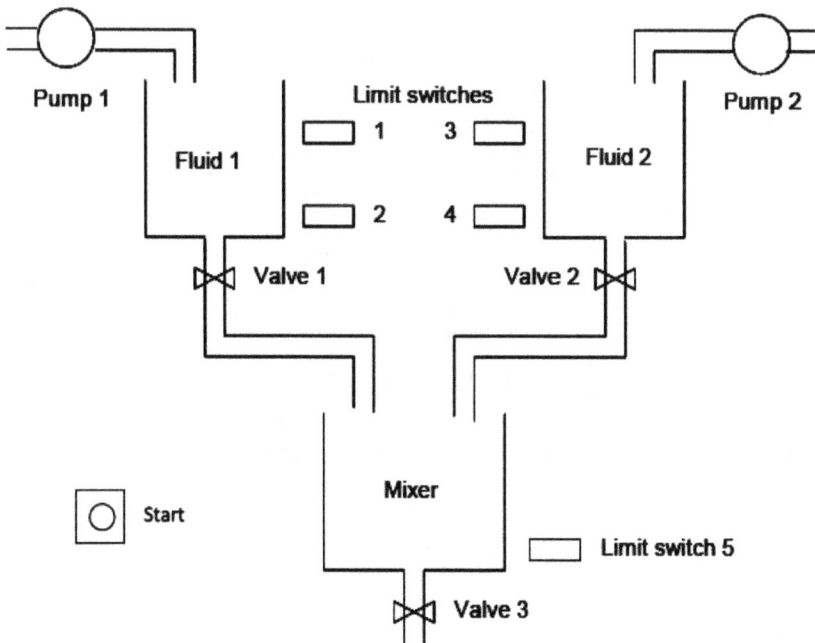

Figure 3.7 System Related to Example 1.

The FSM model for this system is shown in Figure 3.8. The following is the Transition Logic, State Logic and finally the Output Logic of the RLL program.

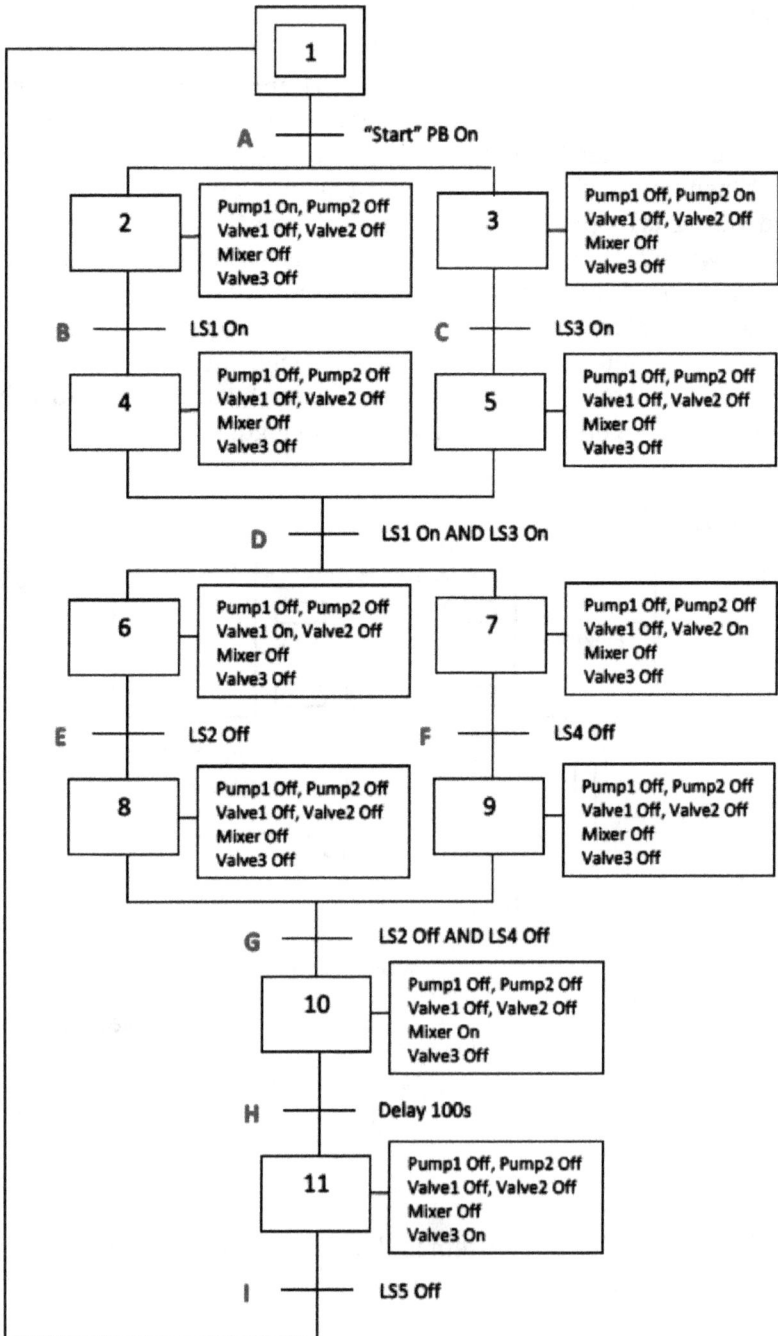

Figure 3.8 The FSM model of Example 1.

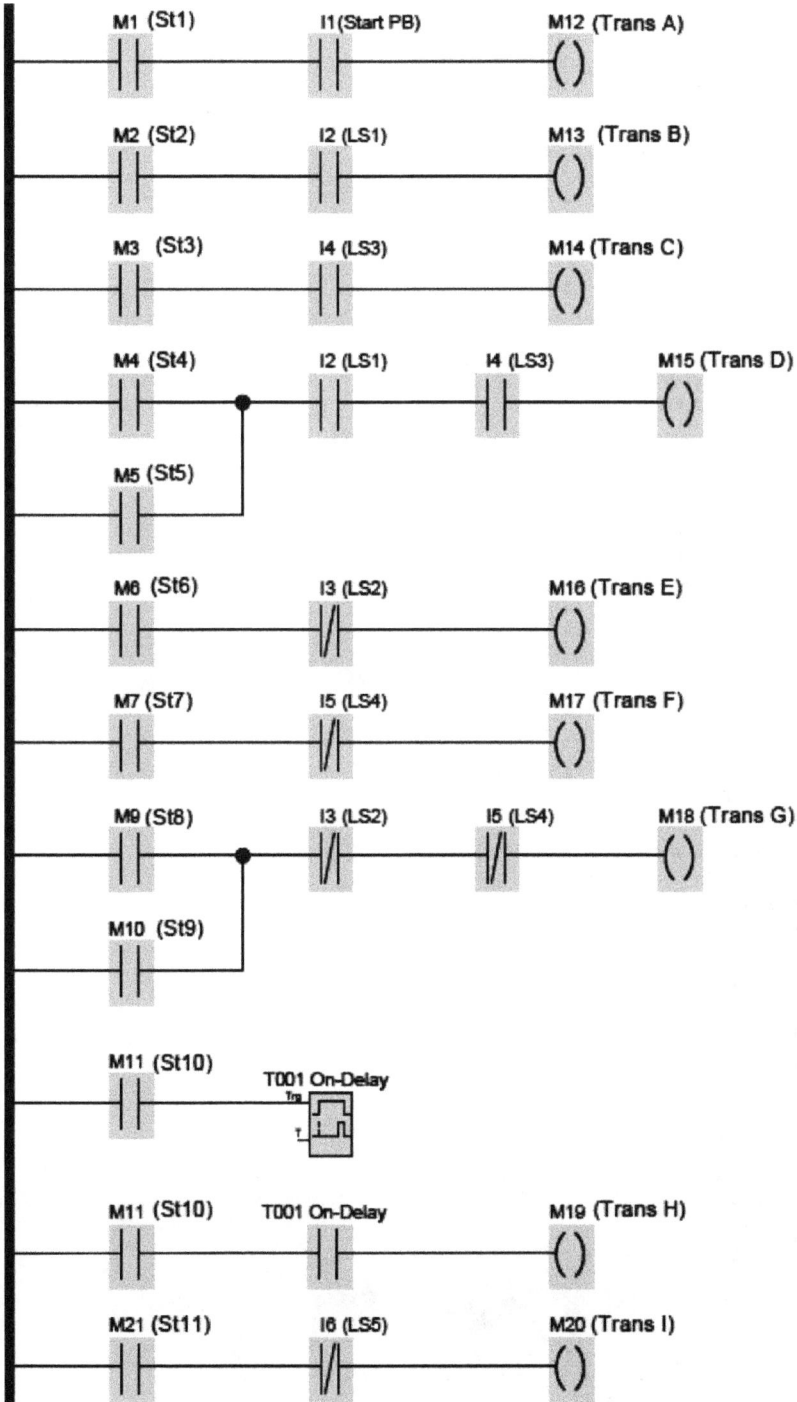

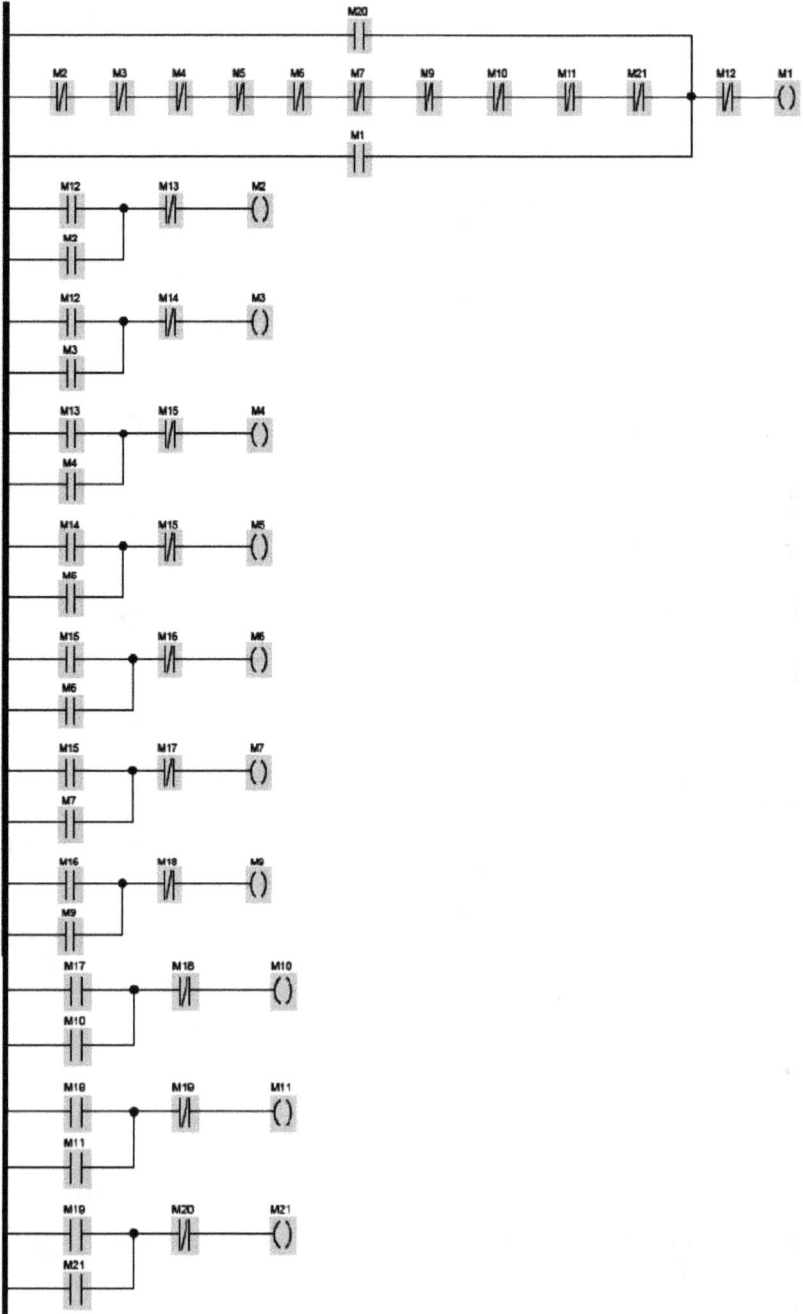

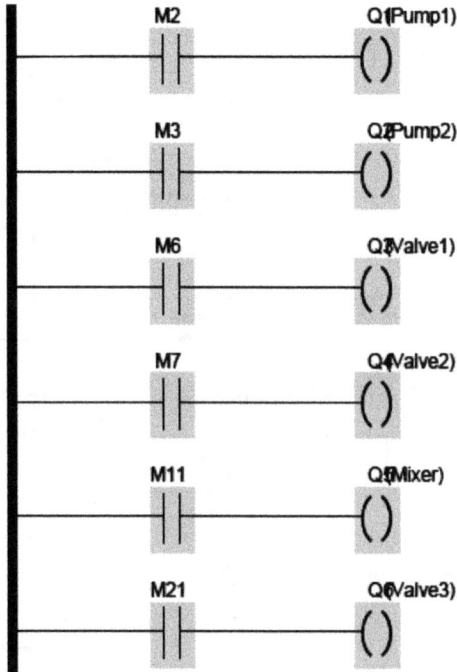

3.9 Example 2

The function of the garage door controller is as follows:
- There is a local push button inside the garage and a remote control push button.
- The door goes up or down when the local push button/remote control push button is pressed once.
- If the local push button/remote control push button is pressed once (while the door is moving), the door stops and if pressed at the second time, it continues to move in the opposite direction.
- Up/Down limit switches are available to stop the door at the bottom and top of the door.
- An optical sensor is located at the bottom in two sides of the door, and if the sensor is activated when the door is coming down (closed), the door is stopped and reverses.
- There is a garage light that illuminates for 5 minutes after the door is opened or closed.

The FSM model for this system is shown in Figure 9.2.

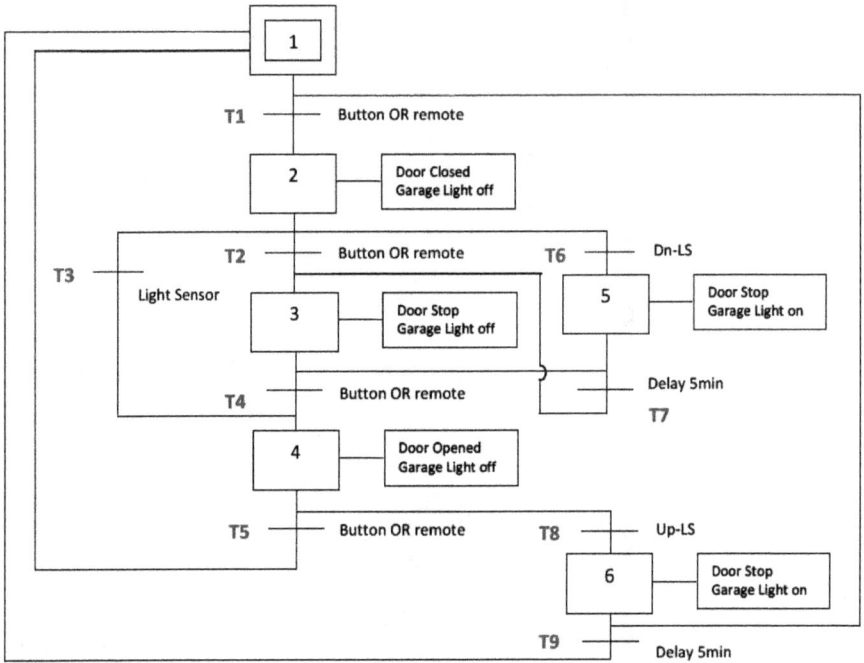

Figure 3.9 FSM model of Example 2.

In the same way as the previous example, we can write Transition Logic, State Logic and finally Output Logic of the RLL program.

In overal, the order of writing the RLL program is as follows:

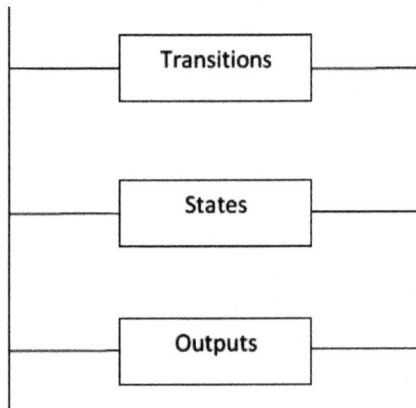

Writing a transition program is such that all the states leading to the specific Transition are considered parallel NO contact states and the inputs cause triggering that transition will be placed as the series NO contacts with the corresponding transition as an output coil in the RLL program.

Note 2: The transitions program does not include self-output contact parallel with the states.

Writing a state program is such that all transitions ending in that state are considered parallel (in NO contact mode) and all transitions leaving the state are considered as NC contacts in series and the desired state will be the corresponding output coil in the RLL program.

Note 3: The state program contains a self-output contact in parallel with input transitions NO contacts.

Note 4: The writing style of the RLL for Idle (Primary state) is slightly different, and its program includes NC contact of other states in series.

Writing the output program is such that in every state the corresponding output occurs is determined and considered as parallel NO contact states connected to that output coil in the RLL program.

Note 5: The output program unlike the state program does not contain a self-output contact in parallel with state NO contacts.

Chapter 4

Application of Petri Nets in PLC Programming

4.1 Introduction

The Petri net (also called P/T network - P: Place, T: Transition) is one of the mathematical modeling languages for extended systems. Petri nets were invented in August 1939 by Carl Adam Petri at the age of 13 to describe chemical processes. A Petri net includes Place, Transition and Arc. Arcs run between Place and Transition and vice versa and never run between Places or Transitions. The places where Arc to Transition is implemented are called Transition Input Places. The places where Arc from a Transition is executed are called Transition Output Places. Graphically, places in a Petri net may include a discrete number of marks called tokens. Distribution of tokens in places forms a marking network. A Transition is fired when sufficient tokens are available in all incoming Arcs. When a Transition fires, it consumes these tokens and sits in all of the output Arcs of those tokens. The firing operation is an uninterruptible process. The implementation of Petri Nets is unclear. Once multiple Transitions are activated at a time, each of them can be fired. If a Transition is enabled, it may or may not fire. Since firing is unclear and several tokens can be anywhere in the network, Petri nets are very useful for modeling the synchronous behavior of large systems. Therefore, it can be said that the FSM model discussed in Chapter 3 is the finite model of Petri nets shown in Figure 4.1.

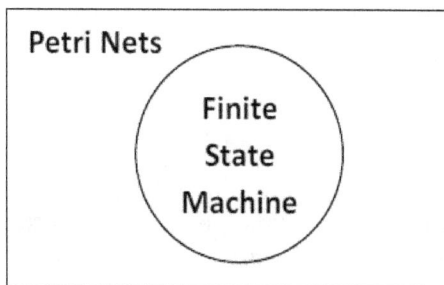

Figure 4.1 Comparison of FSM and Petri nets.

The firing action of a Petri net is illustrated in Figure 4.2. Figure 4.3 shows the conditions under which a Transition cannot fire.

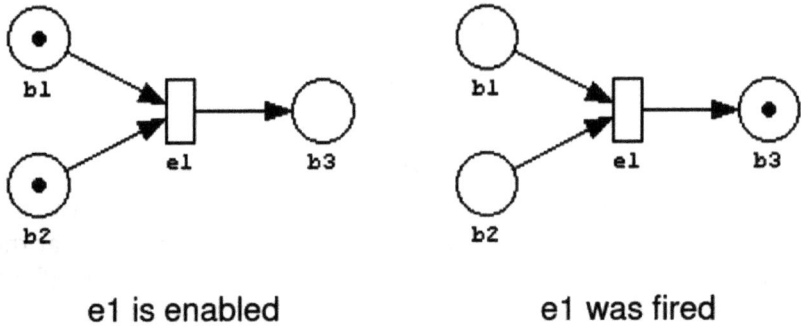

e1 is enabled   e1 was fired

Figure 4.2 The firing operation of a Petri net.

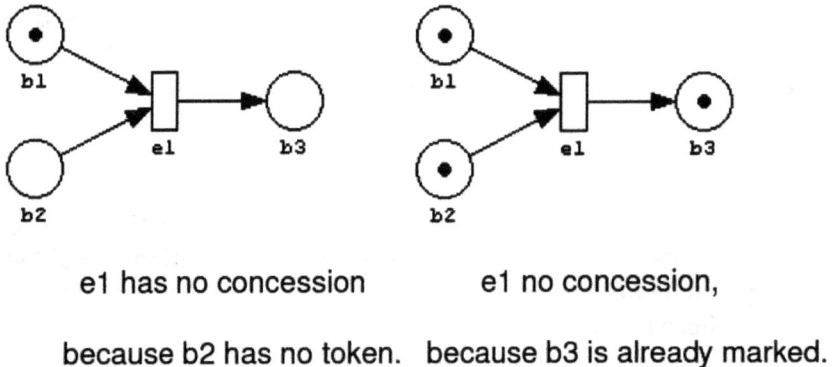

e1 has no concession   e1 no concession,

because b2 has no token.   because b3 is already marked.

Figure 4.3 Conditions for a transition can not to fire.

Petri nets are also called condition-event networks, where the condition is Place and the event is Transition.

Petri nets are suitable for:
- To display non-sequential workflows
- To explicitly show dependency and independence in a set of events
- To model data banks, real-time performance systems, production systems and computer networks
- To represent a system at different levels without changing the tools or methods

In addition to the above features, the simulation of the system by the Petri nets model can confirm the design accuracy, range of performance, reproducibility, viability, availability or inaccessibility of the system modes and features. The main scope of modeling systems with Petri nets is where the events of the system are independent of each other, but their occurrence depends on some constraints (availability of shared resources). There are various software tools available for Petri Nets. We review PetriLLD software in this chapter and explain how to design industrial control issues with this software.

4.2 Introduction to PetriLLD software

PetriLLD is a simple graphics tool used to design PLC applications by creating Petri nets. PetriLLD is built in the form of Petri Nets and has an improved form that only allows one token to be in one place. In PetriLLD, there are external input and output connections that are not common on Petri nets. But its basic ideas are similar to Petri nets. Petri nets are an excellent model for describing the behavior of synchronous systems. Therefore, they are very useful for modeling the behavior of discrete event systems such as production plants in which many operations occur simultaneously. PetriLLD has been developed with automated control systems used in manufacturing plants. PetriLLD is very useful in applications that require a combination of system behaviors simultaneously and sequentially. PetriLLD was developed to produce RLL diagrams for PLCs. The RLL diagram is a graphical language that resembles a series of wired switches. In fact, graphic language is done as a set of Boolean logic commands. By executing and scanning commands, the PLC behaves like real wires and switches in the programming language. Like the RLL diagrams, PetriLLD converts the Petri network into Boolean commands. When Boolean or its equivalent RLL is programmed into the PLC, it will behave the same as the Petri nets, but in this case it will be affected by the actual sensors connected to the PLC and turn PLC operators on and off. Next, the PetriLLD software usage in creating control logic and RLL program is discussed.

4.3 Investigation of an automation problem with PetriLLD

For example, in Figure 4.4, the drill starts at the highest point where the LS1 limit switch is stimulated. When the cycle start button is pressed, the power on lamp illuminates and the clamp is stimulated and the drill goes down (with Head-Z stimulation) and the spindle motor triggered. When reaching the LS2 limit switch, Head-Z is turned off and Head+Z is triggered. Finally, when the LS1 limit switch is activated, the clamp is released and the spindle motor and Head+Z are both switched off. The system stops by pressing the stop button when the drill goes down.

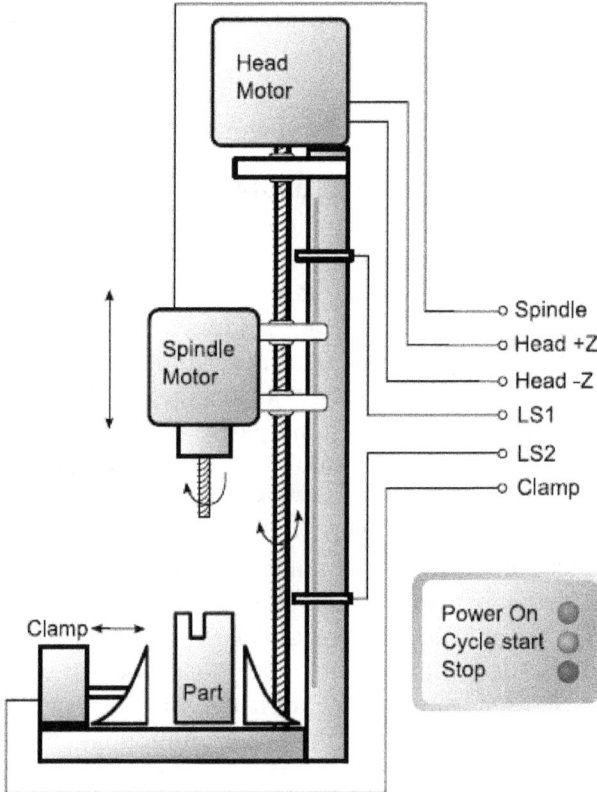

Figure 4.4 Schematic of the drill problem.

Idle is indicated by a circular Place, Inputs with triangular-shaped Places with toggle input only option, Outputs with square-shaped Places with external toggle option and Transitions with green rectangle. Arcs from Idle and Input to Transition and from Transition to the Output is shown in Figure 4.5.

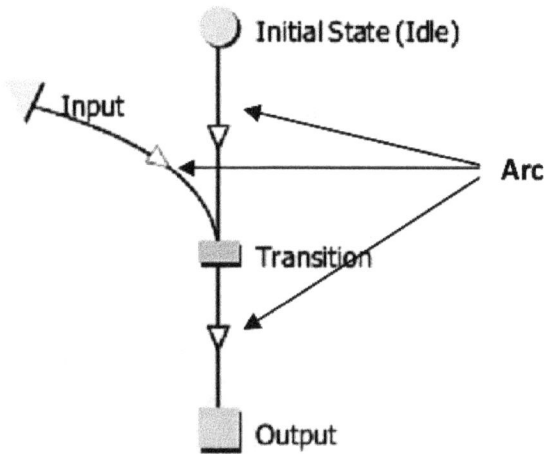

Figure 4.5 PetriLLD Network Elements.

Figure 4.6 shows the solution of the drill control problem with Pe-triLLD software.

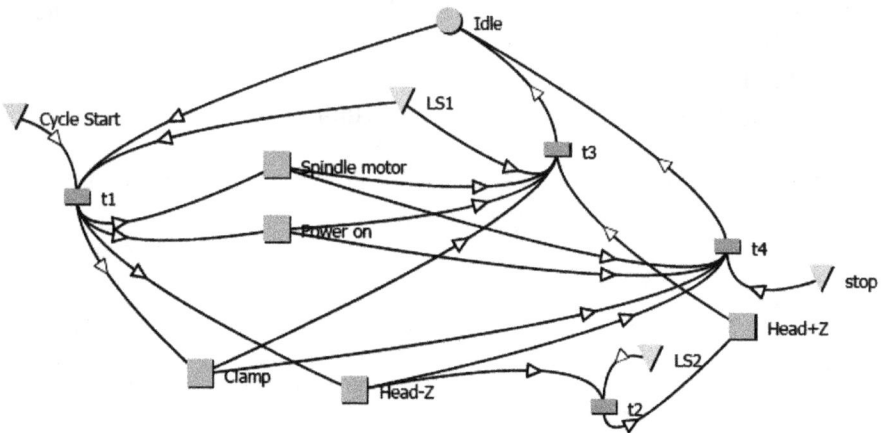

Figure 4.6 Petri net related to the drill control problem.

4.4 Add an instance

Instance is the mapping between the Output and Input places to PLC addresses. It is necessary to add an instance before compiling the Pe-tri net and project simulation. Right-click on the network created and select the new instance option. And we enter the addresses as Table 4.1.

Table 4.1 Addresses used for inputs and outputs

Head-Z	0.00
Head+Z	0.01
Spindle motor	0.02
Clamp	0.03
LS1	0.04
LS2	0.05
Power on	0.06
Cycle Start	0.07
stop	1.00

4.5 Compiling the project

To compile the project, right-click on the project and select Compile project. There are various formats to compile different PLCs, including Omron Cx, Omron CV, Rockwell, IEC61131-3, Java, Visual Basic and Siemens S7. The addresses used in Table 4.1 are suitable for Omron Cx, Omron CV and Rockwell PLCs. If, for example, the Siemens S7 PLC is selected, the instance addresses will be as in Table 4.2.

Table 4.2 Addresses used for inputs and outputs in the Siemens S7

Head-Z	Q0.0
Head+Z	Q0.1
Spindle motor	Q0.2
Clamp	Q0.3
LS1	I0.0
LS2	I0.1
Power on	Q0.4
Cycle Start	I0.2
stop	I0.3

4.6 Project simulation

To simulate the project, right-click on the project and select the simulate project option. A new page is opened and the play button can be pushed by the simulation shown in Figure 4.7.

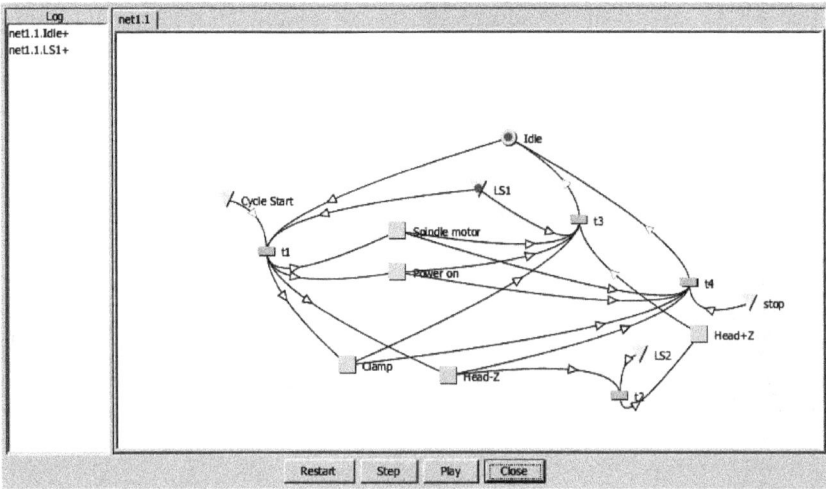

Figure 4.7 Project simulation.

4.7 Converting Petri Net to RLL Program

Since most of the PLCs used are from the Siemens PLC family, so we choose the Siemens S7 compiler to convert the designed Petri net to the RLL program. So, to compile the project, right-click on the project and select Compile project option and in the open screen Siemens S7 AWL (beta) as compile format and click OK. A new window will open and select the name of the out.awl file and press the Write Output button. The message successfully compiled to ... appears. Press the OK button. Open the out.awl file with Notepad. The opened file is a text file and consists of 10 networks (from Network 1 to Network 10) as shown in Table 4.3. Because the software used in Siemens PLC is SIMATIC Step 7 V5.3 and the LD and OLD commands are not defined, we use the A command instead of the LD command and delete the OLD command. If two LD statements are consecutive, the second LD statement becomes O, and at the first of two statements A (...) is used. Also in Network 1, since M0.0 is related to Idle, we only remove the A M0.0 command line from Network 1. In SIMATIC Step 7 V5.3 software we define a project in STL language and in the project OB1 we copy the 10 networks in Table 4.3. Then, using the view: LAD option, we create the corresponding RLL program format shown in Figure 4.8.

Table 4.3 Networks in the out.awl file

Network 1	Network 2	Network 3	Network 4	Network 5	Network 6	Network 7	Network 8	Network 9	Network 10
LD 10.0	LD Q0.0	LD 10.0	LD Q0.4	LD M0.1	LD M0.1	LD M0.0	LD M0.2	LD M0.1	LD M0.1
A M0.0	A 10.1	A Q0.4	A Q0.0	LD Q0.4	LD Q0.0	AN M0.1	LD Q0.1	LD Q0.2	LD Q0.3
A 10.2	AN Q0.1	A Q0.1	A Q0.2	AN M0.4	AN M0.4	LD M0.4	AN M0.3	AN M0.4	AN M0.4
AN Q0.4	= M0.2	A Q0.2	A 10.3	AN M0.3	AN M0.2	O M0.3	OLD	AN M0.3	AN M0.3
AN Q0.0		A Q0.3	A Q0.3	OLD	OLD	OLD	= Q0.1	OLD	OLD
AN Q0.2		AN M0.0	AN M0.0	= Q0.4	= Q0.0	= M0.0		= Q0.2	= Q0.3
AN Q0.3		~ M0.3	AN M0.2						
= M0.1			AN M0.3						
			= M0.4						

Network: 1

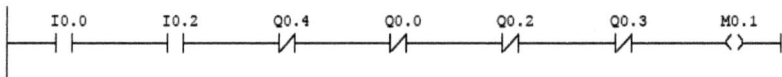

Network: 2

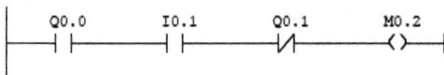

Network: 3

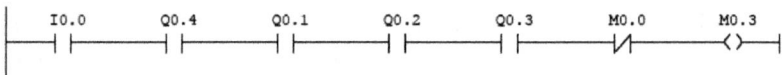

Network: 4

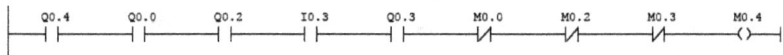

Network: 5

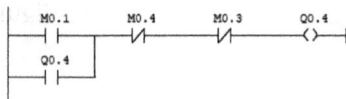

Network: 6

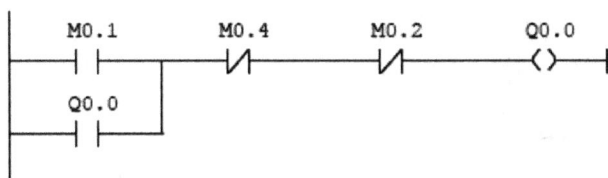

Network: 7

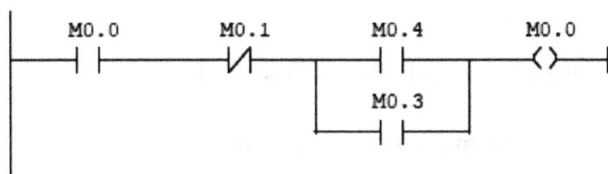

Network: 8

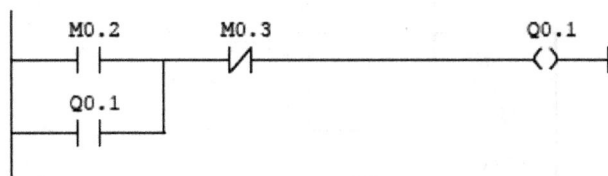

Network: 9

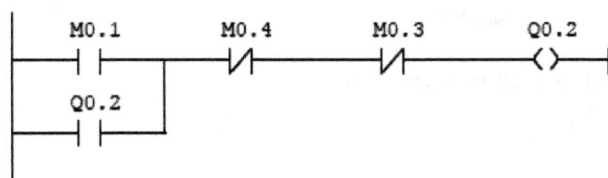

```
Network: 10
```

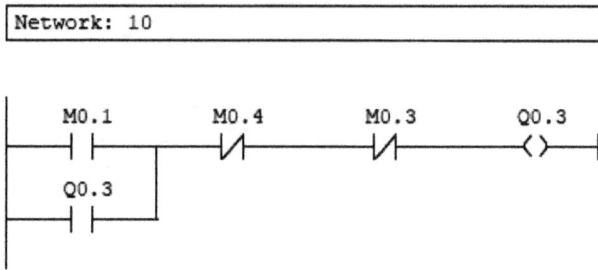

Figure 4.8 RLL program format in Siemens S7.

Let us now consider three examples solved in Chapter 3 by the Petri net method.

4.8 Example 1 (Hydraulic Press Process)

The hydraulic press machine is shown in Figure 4.9.

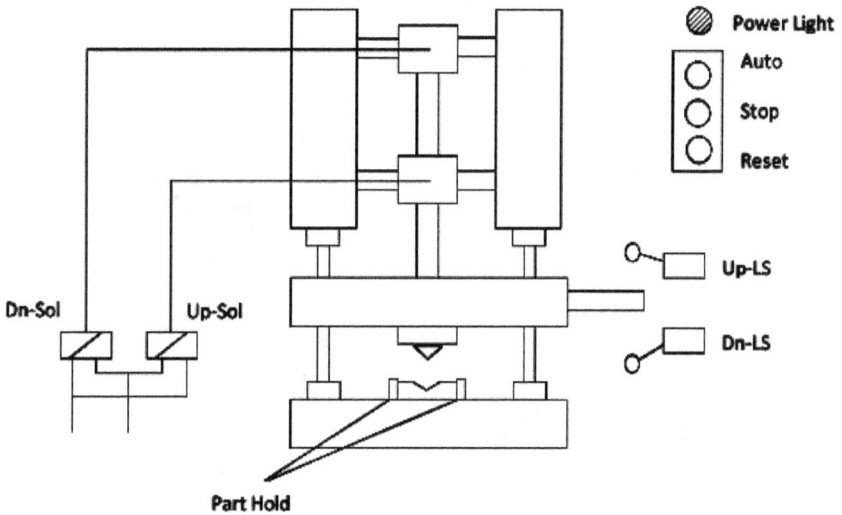

Figure 4.9 Hydraulic press machine.

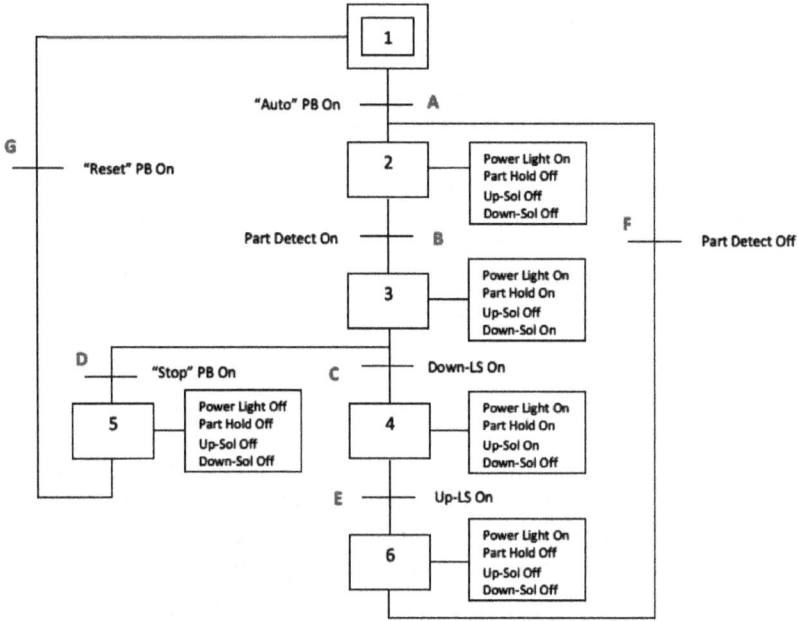

Figure 4.10 FSM Model for Hydraulic Press Process.

The FSM model for the hydraulic press process is shown in Figure 4.10 and the corresponding Petri net in Figure 4.11. The addresses used for inputs and outputs in the Siemens S7 are also listed in Table 4.4.

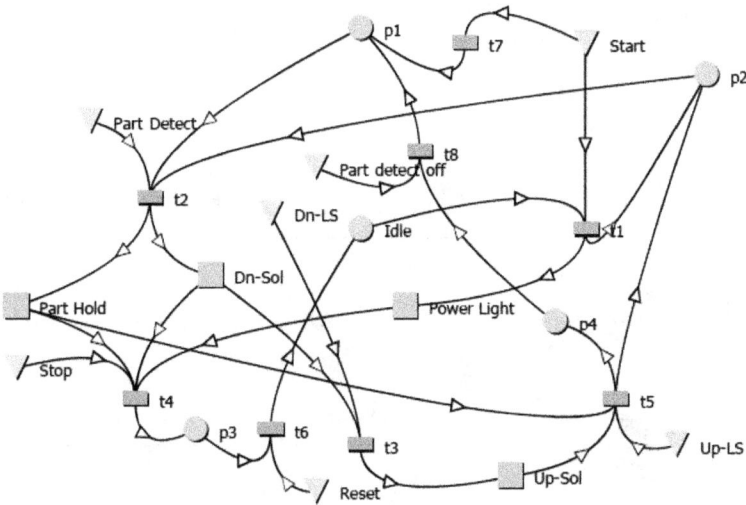

Figure 4.11 Petri net for hydraulic press machine.

Table 4.4 Addresses used for inputs and outputs in the Siemens S7

Part Detect	I0.0
Start	I0.1
Up-LS	I0.2
Dn-LS	I0.3
Stop	I0.4
Reset	I0.5
Part detect off	I0.6
Dn-Sol	Q0.0
Power Light	Q0.1
Up-Sol	Q0.2
Part Hold	Q0.3

Open the out.awl file with Notepad. The opened file is a text file and consists of 18 networks (from Network 1 to Network 18), using the A command instead of LD like the previous example, and removing the OLD command. If two LD statements are consecutive, the second LD statement becomes O, and A (...) statement is used at the first command. Also, in Network 1, since M0.0 is related to Idle, we only remove the A M0.0 command line from Network 1. In SIMATIC Step 7 V5.3 software we define a project in STL language and in OB1 related to the project we overwrite 18 networks in the out.aw file. Then, using the view: LAD option, we create the corresponding RLL program format shown in Figure 4.12.

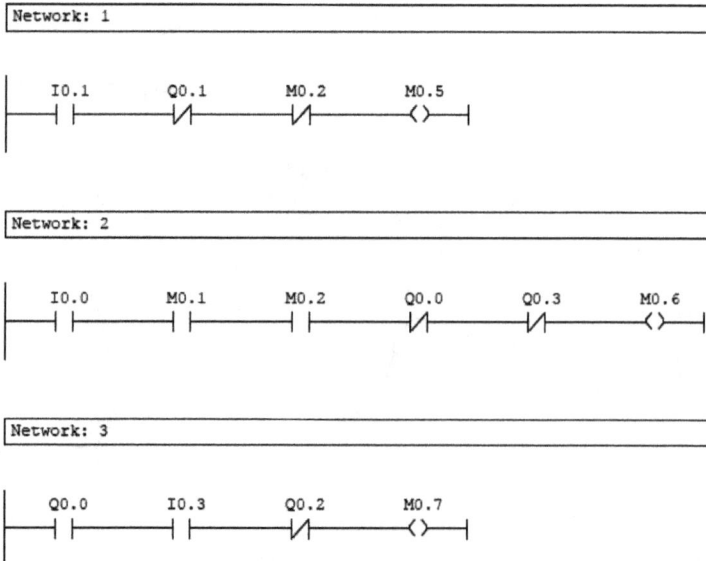

Network: 1

```
    I0.1        Q0.1        M0.2        M0.5
  ──┤ ├────────┤/├─────────┤/├─────────( )──┤
```

Network: 2

```
    I0.0      M0.1      M0.2      Q0.0      Q0.3      M0.6
  ──┤ ├──────┤ ├──────┤ ├──────┤/├──────┤/├──────( )──┤
```

Network: 3

```
    Q0.0        I0.3        Q0.2        M0.7
  ──┤ ├────────┤ ├─────────┤/├─────────( )──┤
```

Network: 4

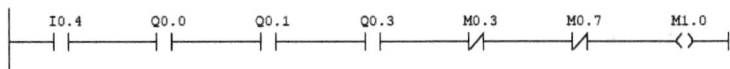

Network: 5

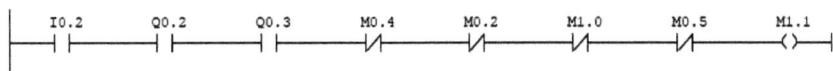

Network: 6

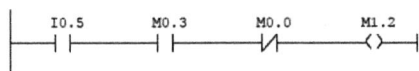

Network: 7

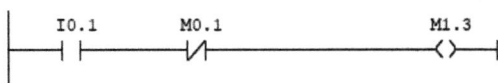

Network: 8

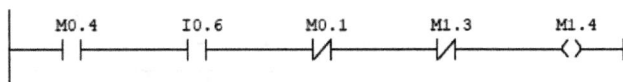

Network: 9

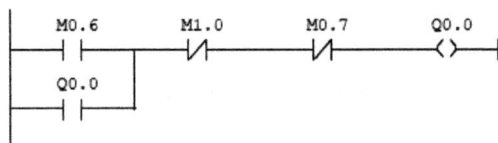

Network: 10

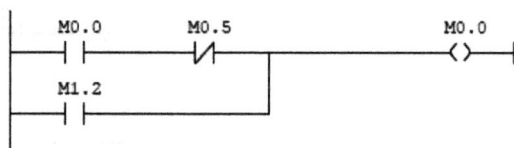

Network: 11

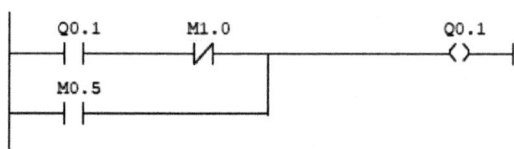

Network: 12

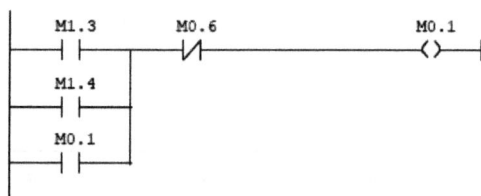

Network: 13

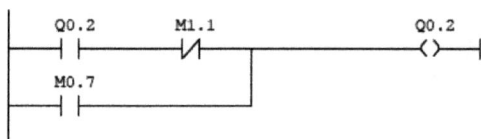

Network: 14

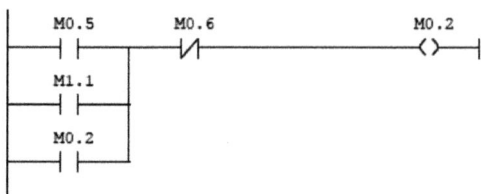

Network: 15

Network: 16

Network: 17

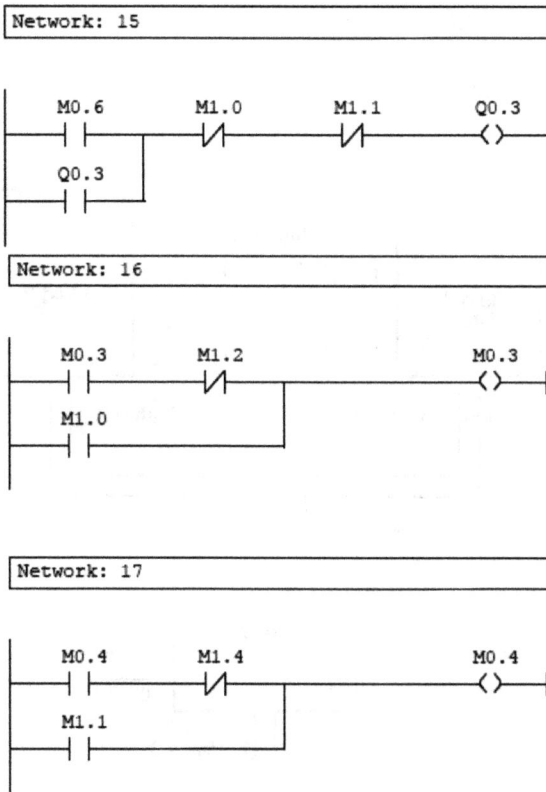

Figure 4.12 RLL application format in Siemens S7.

Note: In the final step, the following application can be used instead of I0.6.

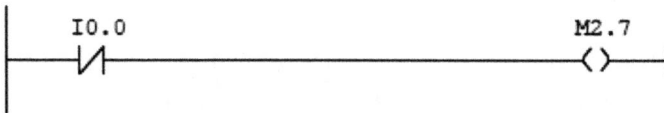

This should be used before Network 8 and instead of I0.6 in Network 8 use M2.7.

4.9 Example 2

The system of Example 2 is illustrated in Figure 4.13.

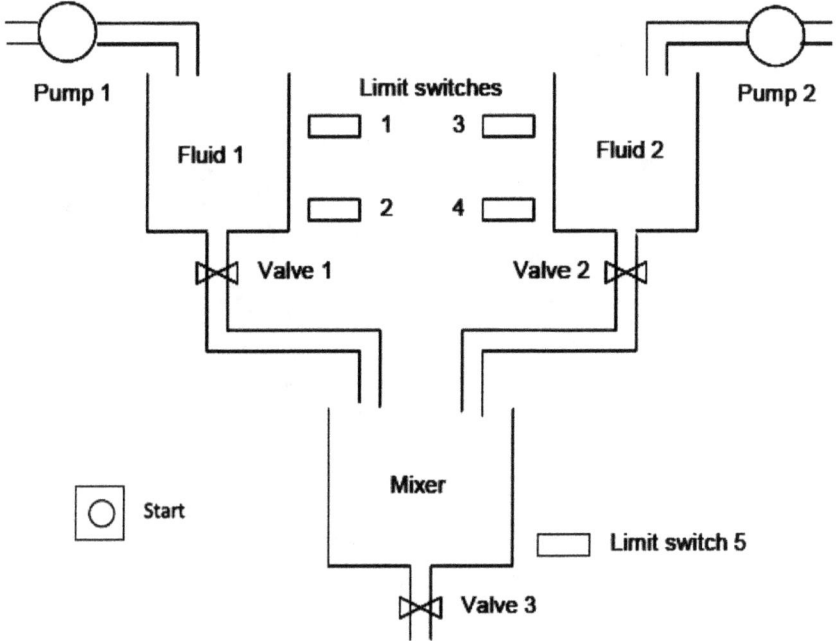

Figure 4.13 System Related to Example 2.

The FSM model for the process of Example 2 is shown in Figure 4.14 and the corresponding Petri net in Figure 4.15. The addresses used for inputs and outputs in the Siemens S7 are also listed in Table 4.5.

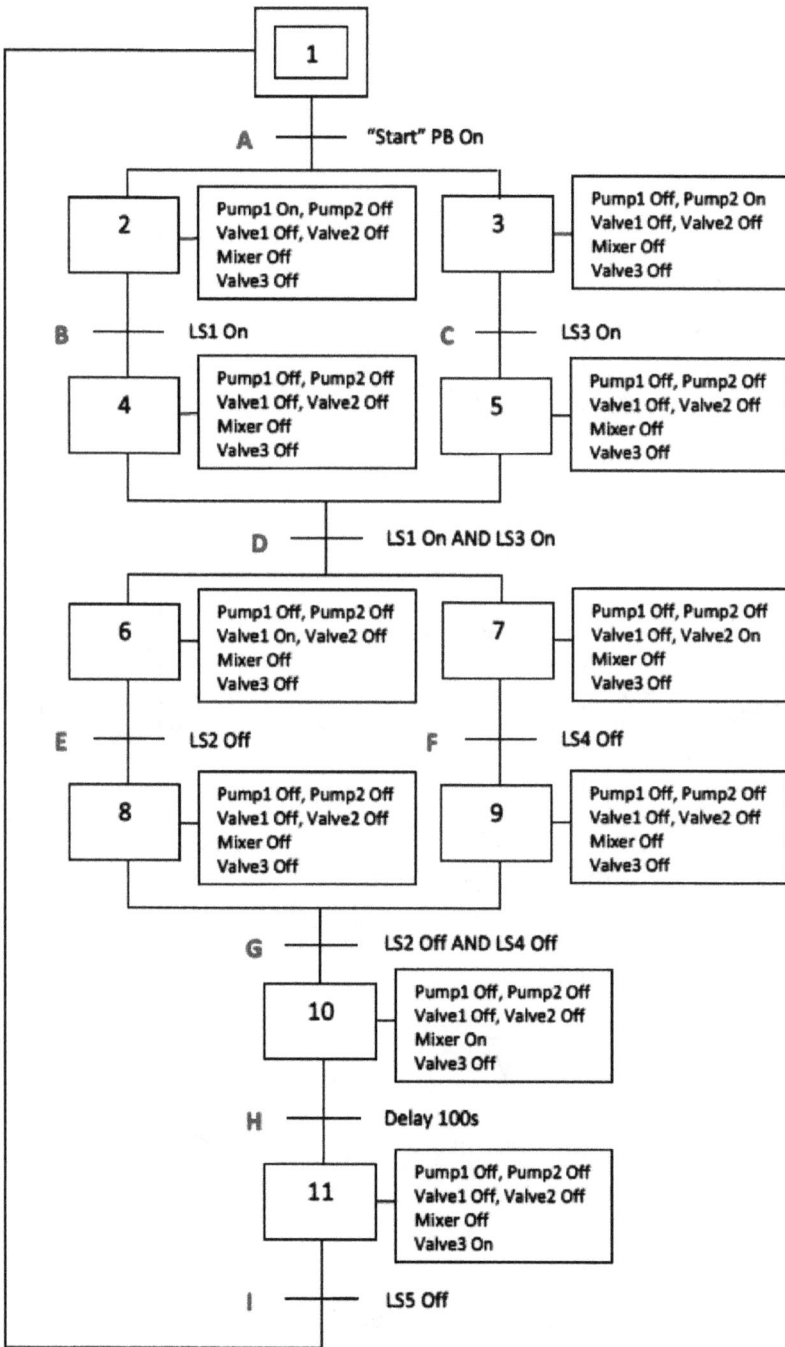

Figure 4.14 Process FSM Model for Example 2.

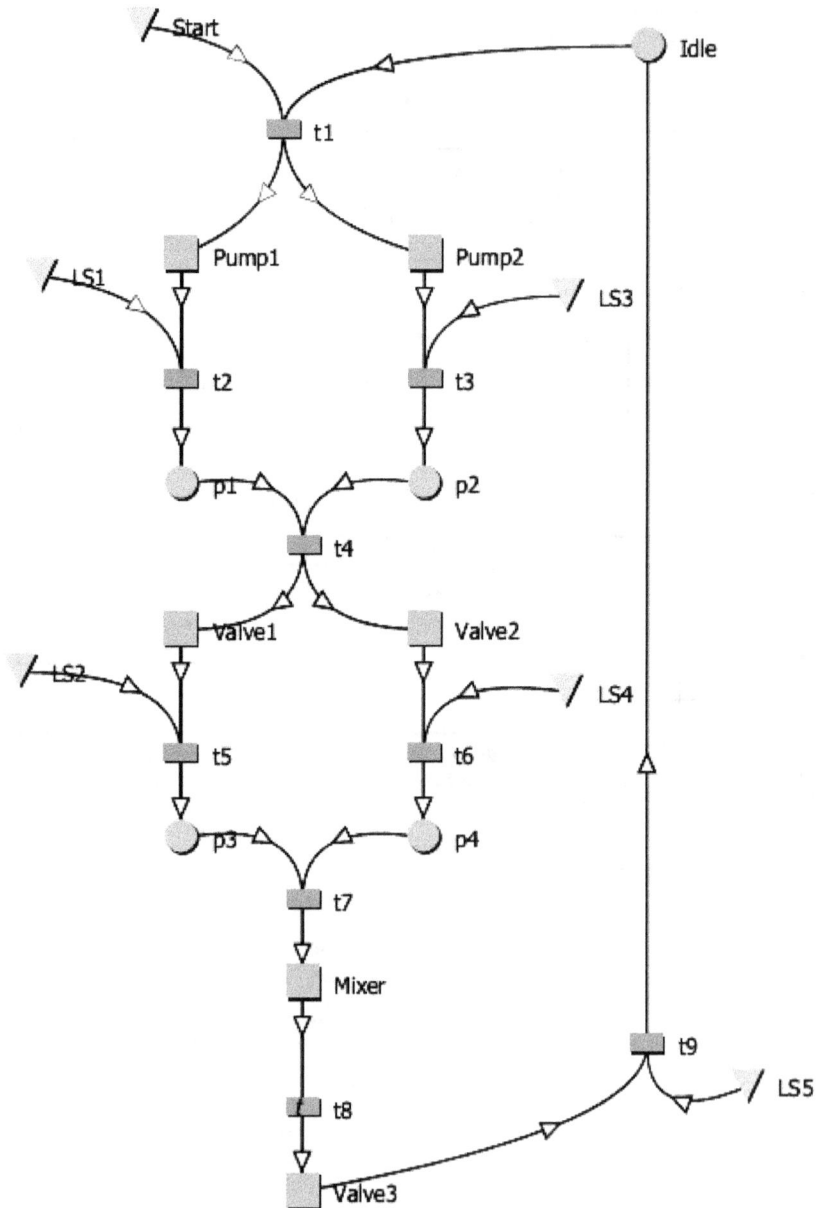

Figure 4.15 Petri net for system Example 2.

Table 4.5 Addresses used for inputs and outputs in the Siemens S7

Start	I0.0
LS3	I0.1
LS4	I0.2
LS5	I0.3
LS1	I0.4
LS2	I0.5
Valve1	Q0.0
Valve3	Q0.1
Valve2	Q0.2
Mixer	Q0.3
Pump1	Q0.4
Pump2	Q0.5

Open the out.awl file with Notepad. The opened file is a text file and consists of 21 networks (from Network 1 to Network 21) which uses the A statement instead of the LD command and we remove the OLD statement. If two LD statements are consecutive, the second LD statement becomes O, and in the first of two statements, A (...) is used. Also in Network 1, since M0.0 is related to Idle state, we only remove the A M0.0 command line from Network 1. And since there's an on delay timer on Network 8, we'll change the program as follows:

TON   T37, +1000

↓

L   S5T#1M40S
SD   T   37
NOP 0
NOP 0
NOP 0
NOP 0

In the SIMATIC Step 7 V5.3 software we define a project in STL language and in the OB1 of the project we overclock 21 networks in the out.aw file. Then, using the view: LAD option, we create the corresponding RLL program format as shown in Figure 4.16.

Network: 1

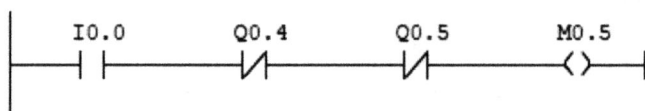

Network: 2

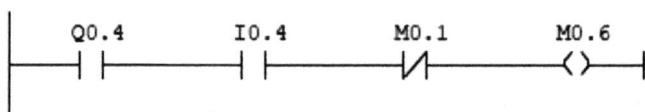

Network: 3

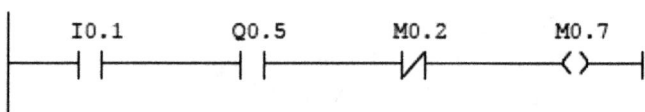

Network: 4

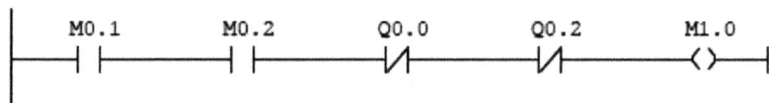

Network: 5

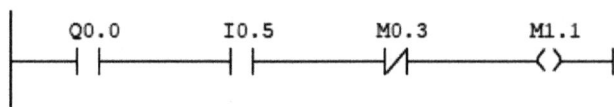

Network: 6

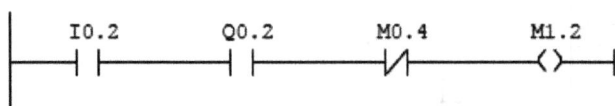

Network: 7

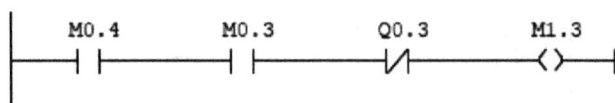

Network: 8

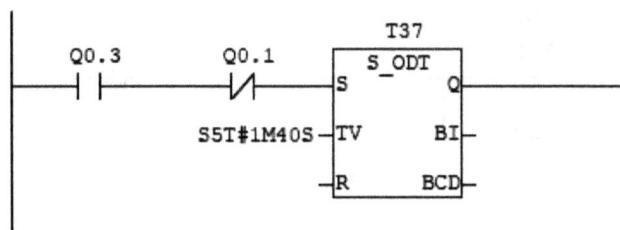

Network: 9

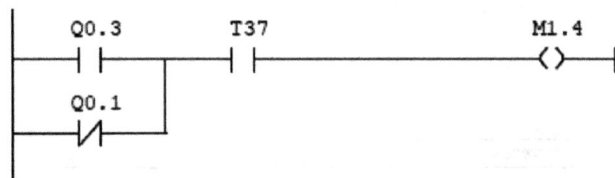

Network: 10

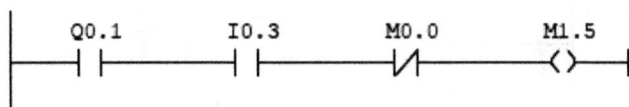

Network: 11

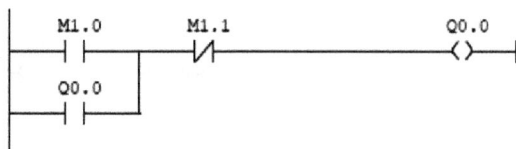

Network: 12

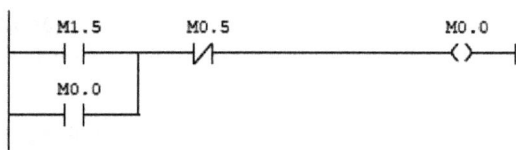

Network: 13

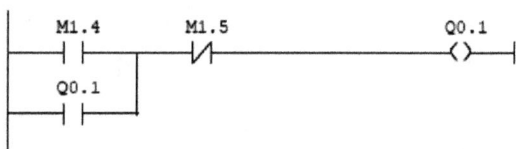

Network: 14

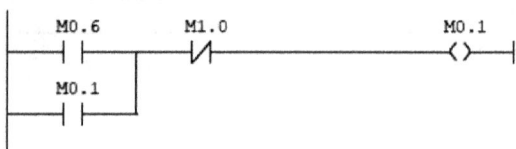

Network: 15

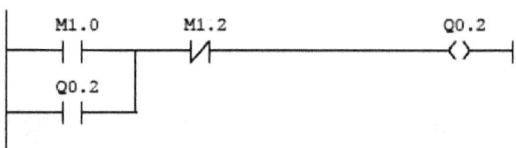

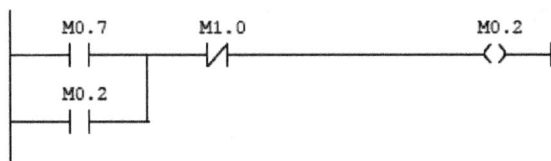

Network: 17

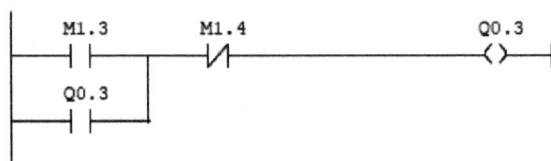

Network: 18

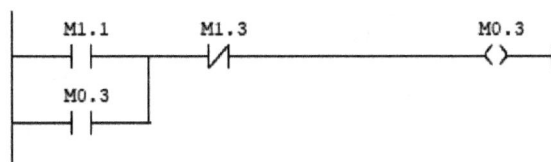

Network: 19

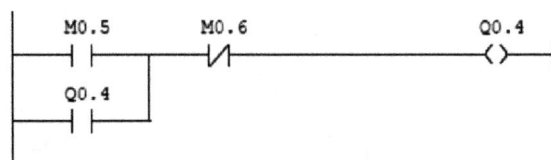

Network: 20

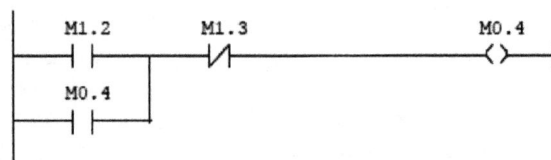

Network: 21

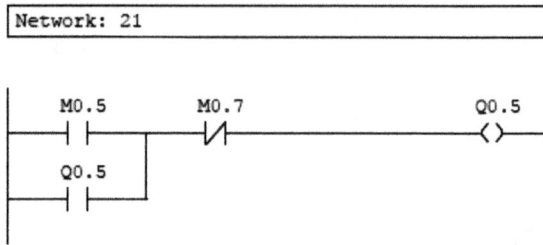

Figure 4.16 RLL program format on Siemens S7.

4.10 Example 3 (Garage Door Control)

The function of the garage door controller is as follows:

- There is a local push button inside the garage and a remote control push button.
- The door goes up or down when the local push button or remote push button is pressed once.

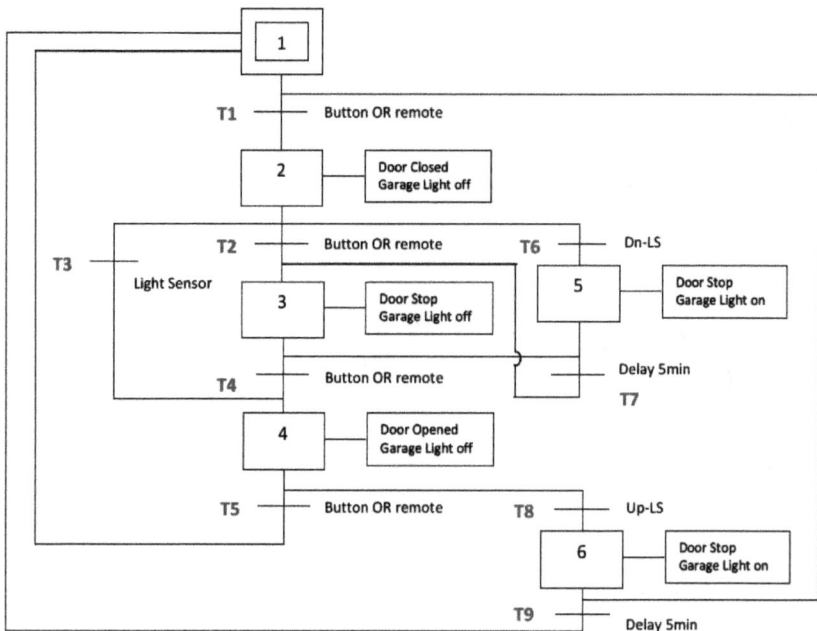

Figure 4.17 FSM model of Example 3.

- If the local push button or remote push button is pressed once (while the door is moving), the door stops and if pressed at a second time, it continues to move in the opposite direction.
- Up/Down switches are available to stop the door at the bottom and top of the door.
- An optical sensor is located at the bottom two sides of the door, and if the sensor is activated when the door is coming down (closed), the door stops and reverses.
- There is a garage light that illuminates for 5 minutes after the door is opened or closed.

The FSM model for this system is shown in Figure 4.17, which is shown earlier in Chapter 3. The corresponding Petri net is shown in Figure 4.18. The addresses used for inputs, outputs, Ts and Ps in the Siemens S7 are also listed in Table 4.6.

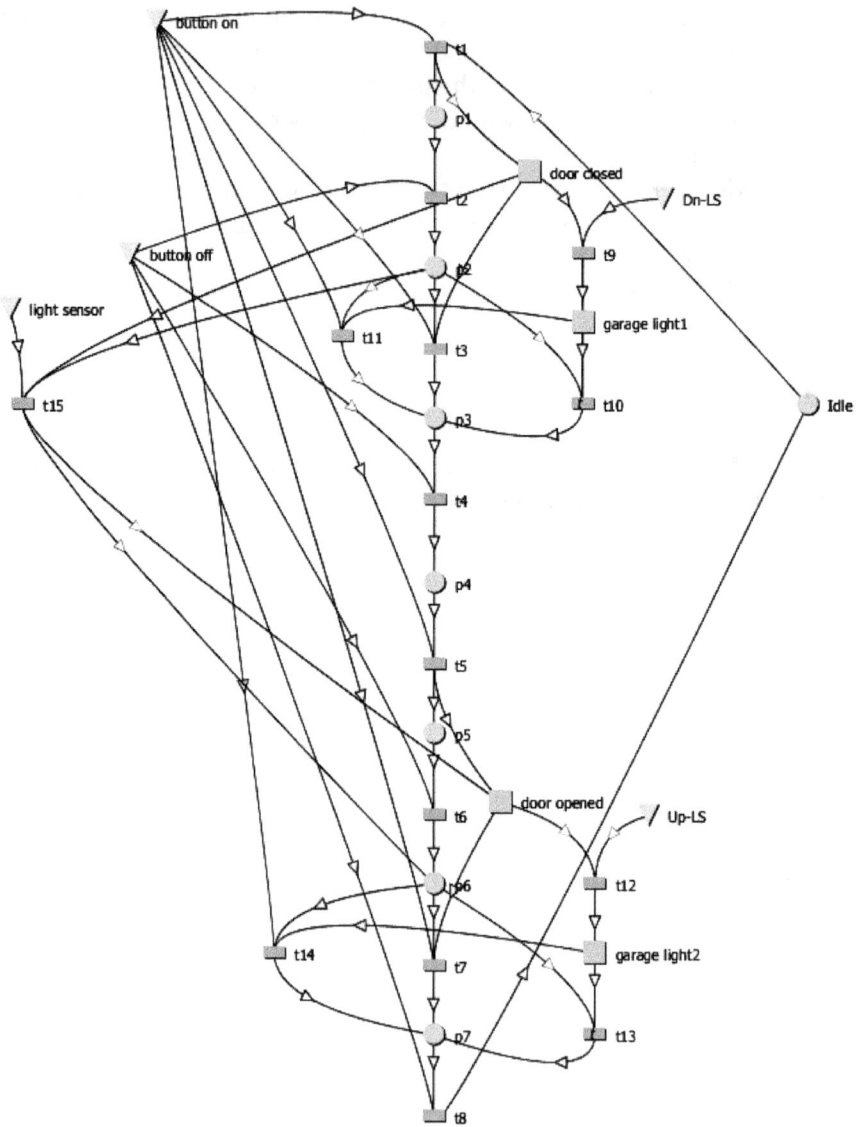

Figure 4.18 Petri Network for System Example 3.

Table 4.6 Addresses used for Inputs, Outputs, Ts and Ps in Siemens S7

Inputs, outputs, Ps and Ts	Addresses
Button	I0.0
Dn-LS	I0.1
Up-LS	I0.2
Light sensor	I0.3
Door close	Q0.0
Door open	Q0.1
Garage light1	Q0.2
Garage light2	Q0.3
Garage light	Q0.4
t1	M0.0
t2	M0.1
t3	M0.2
t4	M0.3
t5	M0.4
t6	M0.5
t7	M0.6
t8	M0.7
t9	M1.0
t10	M1.1
t11	M1.2
t12	M1.3
t13	M1.4
t14	M1.5
t15	M1.6
Idle	M2.0
P1	M2.1
P2	M2.2
P3	M2.3
P4	M2.4
P5	M2.5
P6	M2.6
P7	M2.7

The corresponding RLL program format is shown in Figure 4.19.

Network: 1

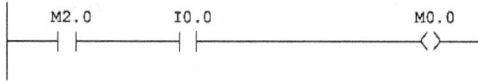

Network: 2

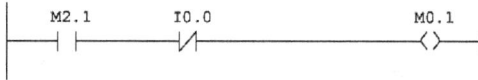

Network: 3

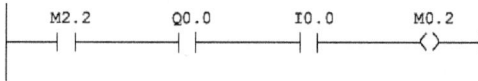

Network: 4

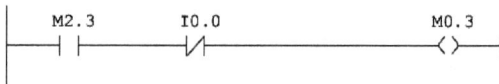

Network: 5

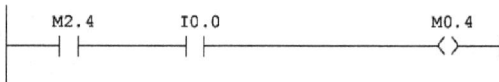

Network: 6

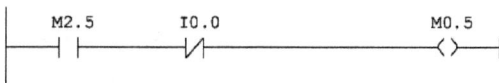

Network: 7

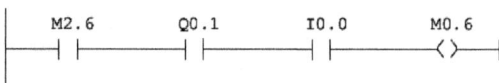

Network: 8

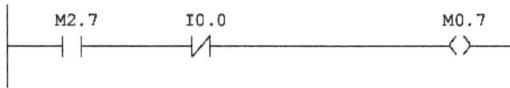

```
      M2.7          I0.0                          M0.7
──┤ ├──────────┤/├──────────────────────────( )──┤
```

Network: 9

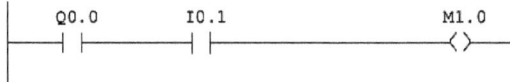

```
      Q0.0          I0.1                          M1.0
──┤ ├──────────┤ ├──────────────────────────( )──┤
```

Network: 10

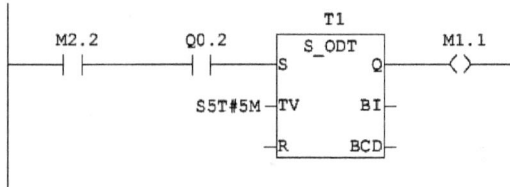

```
                                     T1
      M2.2          Q0.2          ┌─S_ODT─┐       M1.1
──┤ ├──────────┤ ├────────────────S      Q───────( )──┤
                                  │          │
                       S5T#5M─────TV      BI─│
                                  │          │
                                 ─R      BCD─┘
```

Network: 11

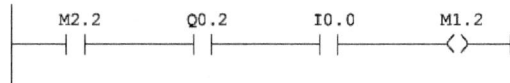

```
      M2.2          Q0.2          I0.0            M1.2
──┤ ├──────────┤ ├──────────┤ ├──────────────( )──┤
```

Network: 12

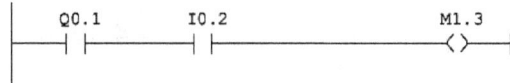

```
      Q0.1          I0.2                          M1.3
──┤ ├──────────┤ ├──────────────────────────( )──┤
```

Network: 13

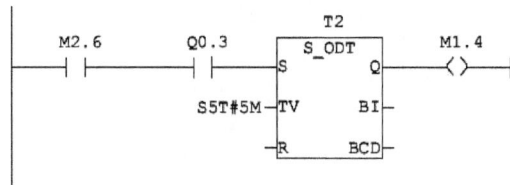

```
                                     T2
      M2.6          Q0.3          ┌─S_ODT─┐       M1.4
──┤ ├──────────┤ ├────────────────S      Q───────( )──┤
                                  │          │
                       S5T#5M─────TV      BI─│
                                  │          │
                                 ─R      BCD─┘
```

Network: 14

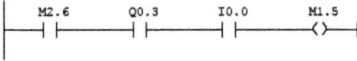
```
     M2.6        Q0.3        I0.0        M1.5
├───┤ ├────────┤ ├────────┤ ├────────( )───┤
```

Network: 15

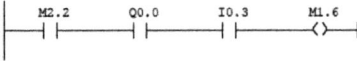
```
     M2.2        Q0.0        I0.3        M1.6
├───┤ ├────────┤ ├────────┤ ├────────( )───┤
```

Network: 16

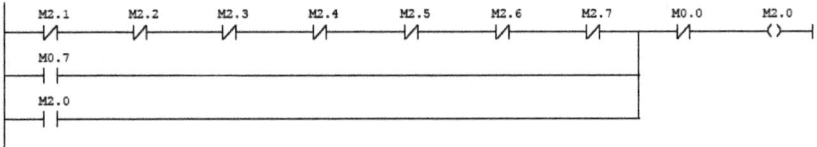
```
     M2.1    M2.2    M2.3    M2.4    M2.5    M2.6    M2.7    M0.0    M2.0
├────┤/├─────┤/├─────┤/├─────┤/├─────┤/├─────┤/├─────┤/├─────┤/├────( )──┤
│
│    M0.7
├────┤ ├─
│
│    M2.0
├────┤ ├─
```

Network: 17

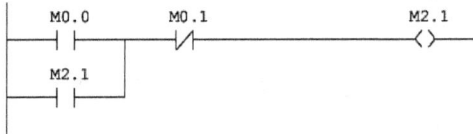
```
        M0.0        M0.1                    M2.1
├───────┤ ├─────────┤/├─────────────────( )───┤
│
│       M2.1
├───────┤ ├─
```

Network: 18

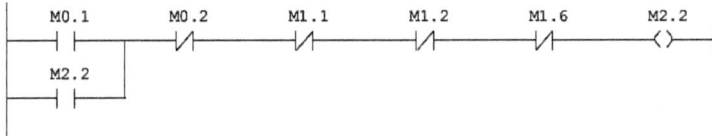
```
        M0.1        M0.2        M1.1        M1.2        M1.6        M2.2
├───────┤ ├─────────┤/├─────────┤/├─────────┤/├─────────┤/├──────( )───┤
│
│       M2.2
├───────┤ ├─
```

Network: 19

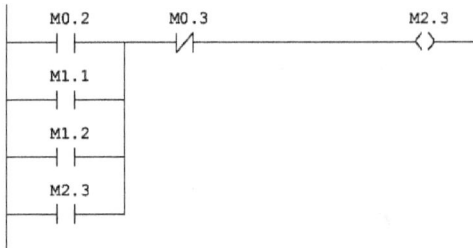
```
        M0.2        M0.3                    M2.3
├───────┤ ├─────────┤/├─────────────────( )───┤
│
│       M1.1
├───────┤ ├─
│
│       M1.2
├───────┤ ├─
│
│       M2.3
├───────┤ ├─
```

Network: 20

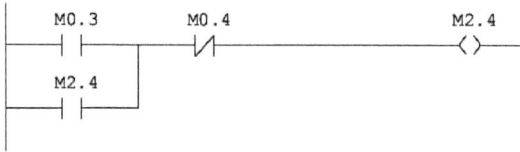

Network: 21

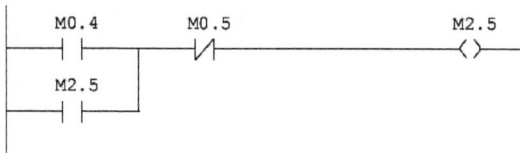

Network: 22

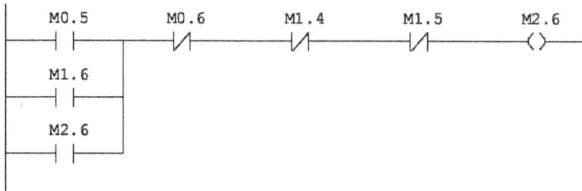

Network: 23

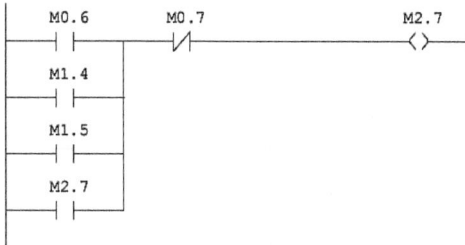

Network: 24

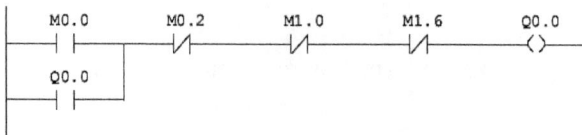

Network: 25

Network: 26

Network: 27

Network: 28

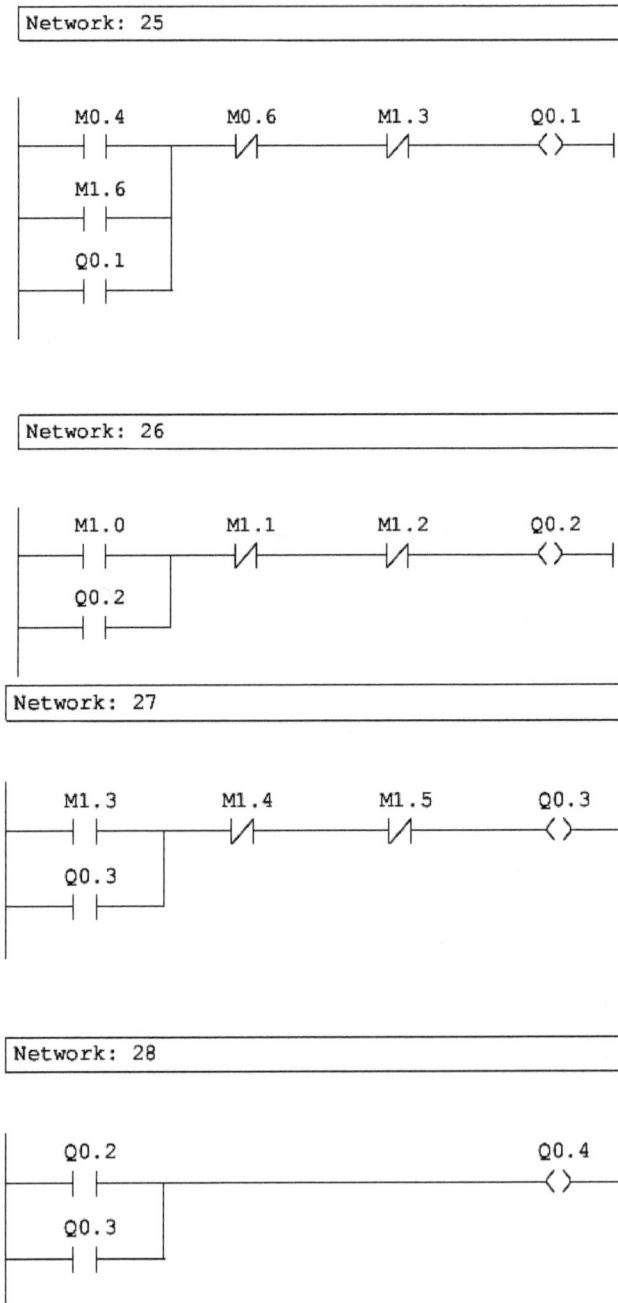

Figure 4.19 RLL program format on the Siemens S7.

Note 1: Remote input can be considered as Remote: I0.4 and be used parallel to I0.0: Button in NO contact mode of Button (Remote is also in NO mode) and series with Button in NC contact mode (Remote is also in NC mode) which has not been shown in Figure 4.19. The output of Garage light: Q0.4 also comes from the parallelization of the two outputs of Garage light1: Q0.2 and Garage light2: Q0.3, as shown in Network 28 of Figure 4.19.

The order of writing the RLL program in Figure 4.19 is as follows:

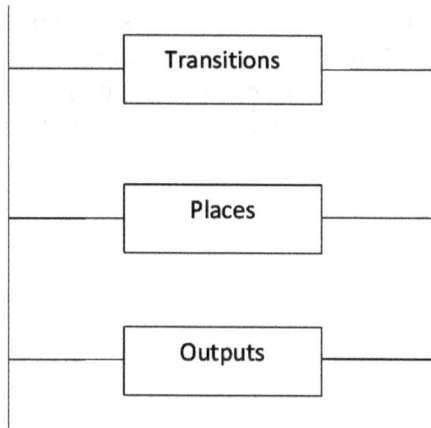

Writing a transition program is such that all the places leading to the specific Transition are considered parallel NO contact places and the inputs cause triggering that transition will be located as the series NO contacts with the corresponding transition as an output coil in the RLL program.

Note 2: The transitions program does not include self-output contact parallel with the places.

Writing a place program is such that all transitions ending in that place are considered parallel (in NO contact mode) and all transitions leaving the place are considered as NC contacts in series and the desired place will be the corresponding output coil in the RLL program.

Note 3: The place program contains a self-output contact in parallel with input transitions NO contacts.

Note 4: The writing style of the RLL for Idle (Primary place) is slightly different, and its program includes NC contact of other places in series.

Writing the output program is exactly like the place program.

Note 5: The output program like the state program contains a self-output contact in parallel with transition NO contacts.

Note 6: In this particular program, using the PetriLLD software compiler (out.awl file), unlike the previous examples, does not cause in a correct result. Therefore, in cases where simulation with the out.awl file does not result in a correct result, the corresponding RLL program can be easily obtained from the created Petri net using the method described above.

Chapter 5

PLC Hardware Components

5.1 Introduction

This chapter discusses various PLC hardware such as I/O, CPUs and memory, contactors, manual switches, mechanical switches, different type of sensors and output control devices.

5.2 I/Os

Figure 5.1 shows the rack-based I/O modules. I/O modules are interface modules that receive signals from process machines or devices and convert them into signals that can be used by the controller.

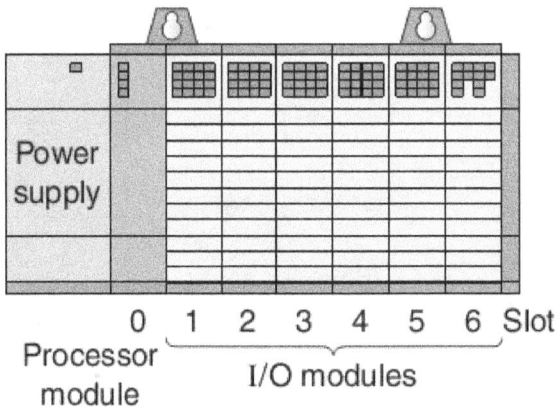

Figure 5.1 Rack-based I/O modules.

A logic rack consists of 128 inputs and 128 outputs. In other words, a logical rack consists of 8 input words and 8 output words. Therefore, each logic rack can consist of 8 groups of I/O (from 0 to 7 or 128 discrete I/Os). One advantage of a PLC system is the ability to place I/O modules near the device (in the field) to minimize the amount of wiring as shown in Figure 5.2. The processor receives the signals from the remote input modules through the communication modules and returns the output signals through the communication modules.

A rack is called a remote rack when it is away from the processor module.

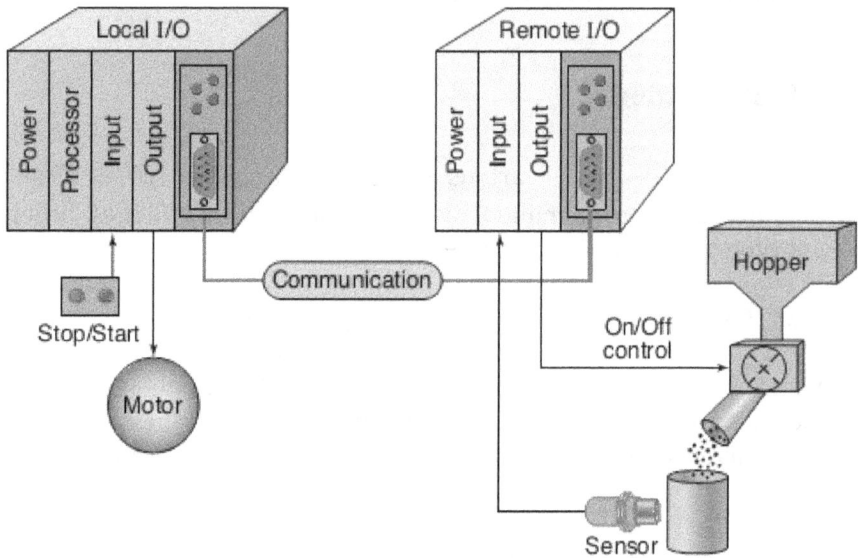

Figure 5.2 Remote I/O rack.

5.3 Discrete I/O modules

The most common type of I/O module is the discrete I/O module (Figure 5.3). These types of modules are connected to input devices of the nature of ON/OFF such as selector switch, push button and limit switch. Similarly, output controls are limited to devices such as lights, relays, solenoids and motor starters, which require simple ON/OFF switching. Each discrete I/O module is powered by voltage sources with different AC and DC amplitudes as listed in Table 5.1.

Table 5.1 Different voltage values for discrete I/O modules

Input Interfaces	Output Interfaces
12 V **AC/DC** /24 V **AC/DC**	12–48 V **AC**
48 V **AC/DC**	120 V **AC**
120 V **AC/DC**	230 V **AC**
230 V **AC/DC**	120 V **DC**
5 V **DC** (TTL level)	230 V **DC**
	5 V **DC** (TTL level)
	24 V **DC**

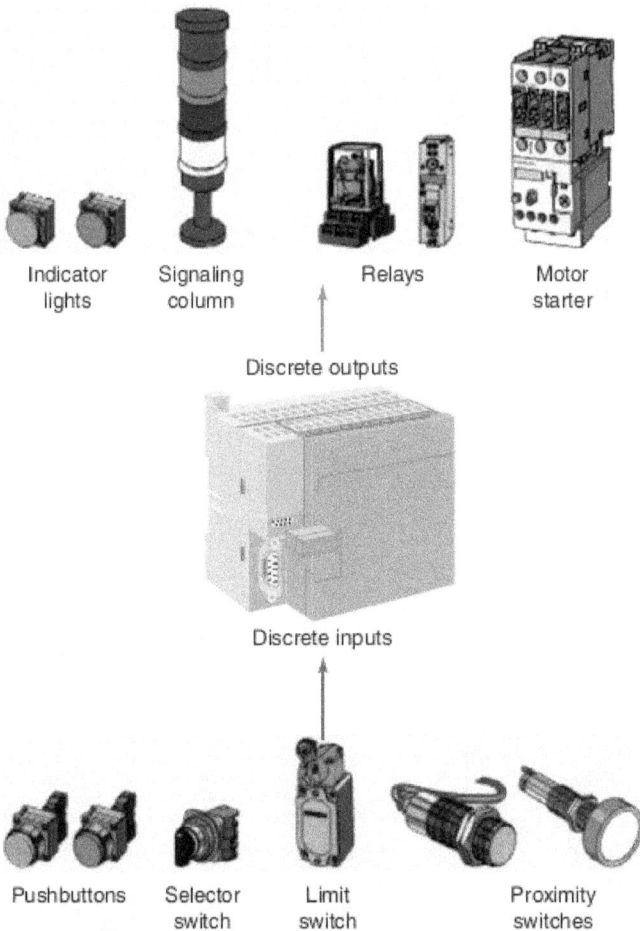

Indicator   Signaling   Relays   Motor
lights   column   starter

Discrete outputs

Discrete inputs

Pushbuttons   Selector   Limit   Proximity
               switch   switch   switches

Figure 5.3 Discrete input and output devices.

5.4 Analog I/O modules

Discrete input and output devices that have only two modes on and off. By analogy, analog devices represent physical quantities that can have infinite quantities. Sample analog inputs and outputs vary from 0 to 20 mA, 4 to 20 mA or 0 to 10 V. Figure 5.4 shows how analog input and output PLC modules are used to measure and display the level. Analog input modules typically have multiple input channels that allow 4, 8, or 16 devices to be connected to the PLC. Two basic types of analog input modules are voltage-sensing and current-sensing. The ring power supply may be supplied by the sensor or may be supplied by the analog output module as shown in Figure 5.5.

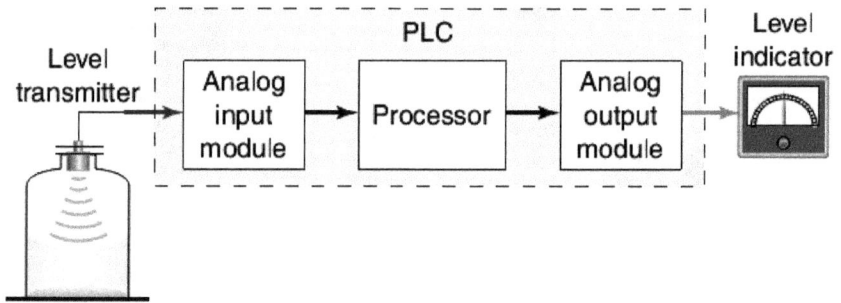

Figure 5.4 PLC Input and Output Analog Modules.

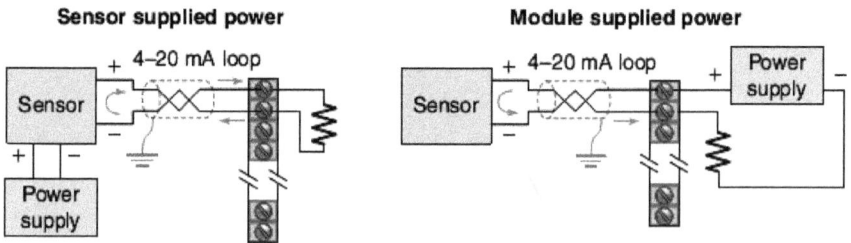

Figure 5.5 Power Supply by Sensor and Analog Module.

Shielded twisted pair cables are typically recommended for connecting any type of analog input signal. Figure 5.6 illustrates how to use analog I/O modules in a PLC control system. In this system, the PLC controls the amount of fluid in the tank by adjusting the percentage of valve opening.

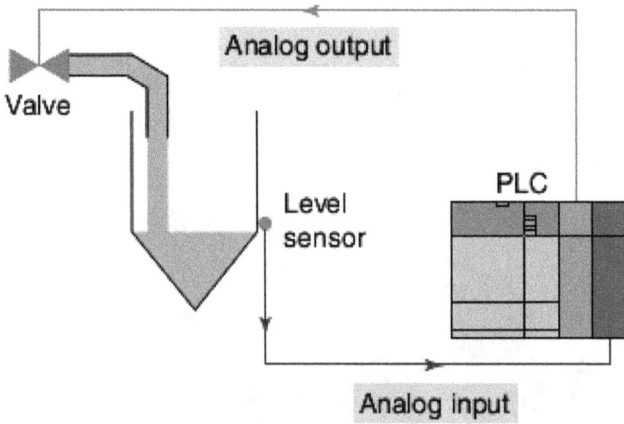

Figure 5.6 Analog I/O Control System.

5.5 Specific I/O modules

a) High speed counter module

The high speed counter module is used in applications that require speeds higher than the RLL program capability. The High Speed Counter Module (Figure 5.7) is used for pulse counting of sensors, encoders, and switches operating at high speeds, which has the electronics required for CPU-independent counting and their counting speeds are from zero to 100 kHz.

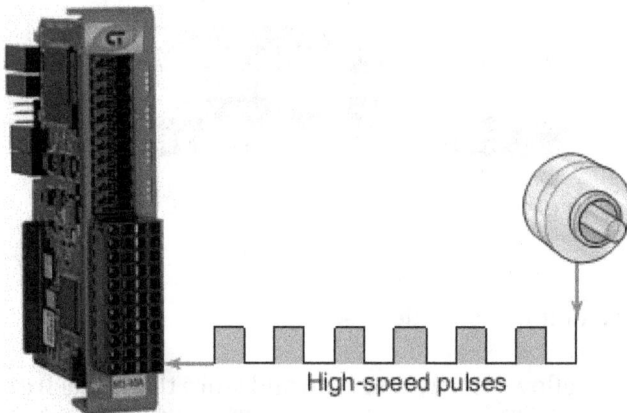

Figure 5.7 High Speed Counter Module.

b) THUMBWHEEL module

This module allows THUMBWHEEL switches (Figure 5.8) to be connected to the PLC which is used in PLC control programs.

Figure 5.8 THUMBWHEEL switch.

c) TTL module

The TTL module (Figure 5.9) allows the transmission and reception of TTL signals. This module allows the devices that generate the TTL signal to communicate with the PLC processor.

Figure 5.9 TTL module.

d) Encoder-Counter module

This module allows the user to read and store the signal from the encoder (Figure 5.10) on a real-time basis. This information can be read later by the processor.

Figure 5.10 Encoder.

e) BASIC or ASCII module

The ASCII Core Module (Figure 5.11) runs user-written BASIC and C programs. These applications are independent of the PLC processor and allow easy and fast communication between external remote devices and the PLC processor. Uses of this module include: Communication with bar code readers, robots, printers and displays.

Figure 5.11 BASIC module.

f) Stepper motor module

The stepper motor module provides the pulse train to the motor (Figure 5.12) and thus controls it. The commands for this module are specified in the PLC control program.

Figure 5.12 Stepper motor.

g) BCD output module

The BCD output module enables the PLC to operate devices that require a BCD encoded signal such as a seven segment (Figure 5.13).

Figure 5.13 Display of the seven Segment.

Note: Some special modules are called smart I/O modules because they have microprocessors on board that can work in parallel with PLCs, which include:

- PID module

The PID module (Figure 5.14) is used in process control applications that use PID algorithms. This is a complex programming algorithm based on mathematical calculations. The PID module allows process control to be performed outside the CPU. This way the CPU is safe from complex computing. The main function of this module is to provide control operations for setting process variables such as temperature, flow, level or speed within a set point.

Figure 5.14 PID module.

- Motion and position control module

Motion and position control modules are used in various applications related to high speed precision machining and packaging operations. The intelligent positioning and motion control modules allow the PLC to control the stepper motor and servo motor. These systems require a drive that includes power electronics converting signals from the PLC module to the required motor signals (Figure 5.15).

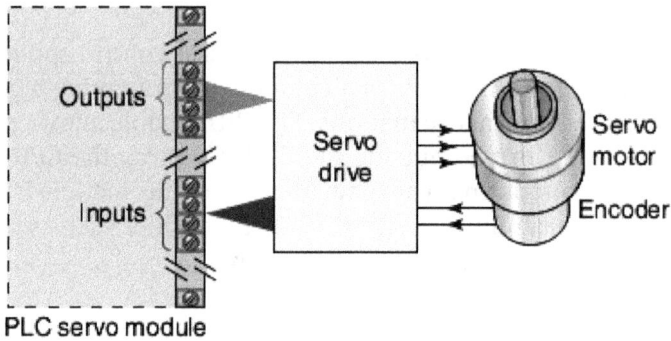

Figure 5.15 Servo module of PLC.

5.6 Telecommunication modules

Serial telecommunication modules (Figure 5.16) are used to exchange point-to-point connections with other intelligent devices. Such connections typically occur with computers, operator stations, process control systems, and other PLCs.

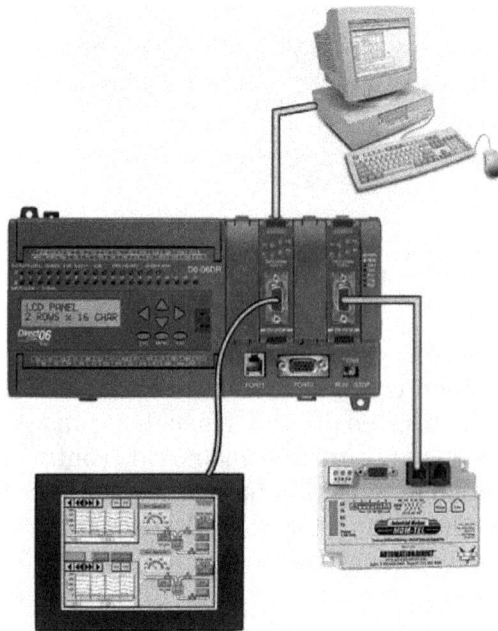

Figure 5.16 Serial Telecommunication Module.

The telecommunication modules allow the user to connect the PLC to high-speed local area networks, which may be different from the telecommunications network created by the PLC.

5.7 Central Processing Unit (CPU)

CPU, controller, and processor are all terms used by different module makers for the CPU. Processors differ in processing speed and memory. A CPU module is divided into two parts: the CPU section and the memory section (Figure 5.17). The CPU executes the program and makes the necessary decisions by the PLC to operate and communicate with other modules. The memory section electronically stores the PLC program along with other digital information. The PLC power supply provides the necessary power (usually 5 VDCs) for the processor and I/O modules with their socket located on the rear panel of the rack (Figure 5.18). The CPU runs the operating system, manages the memory, monitors the inputs, evaluates the user's RLL program, and converts it into appropriate outputs.

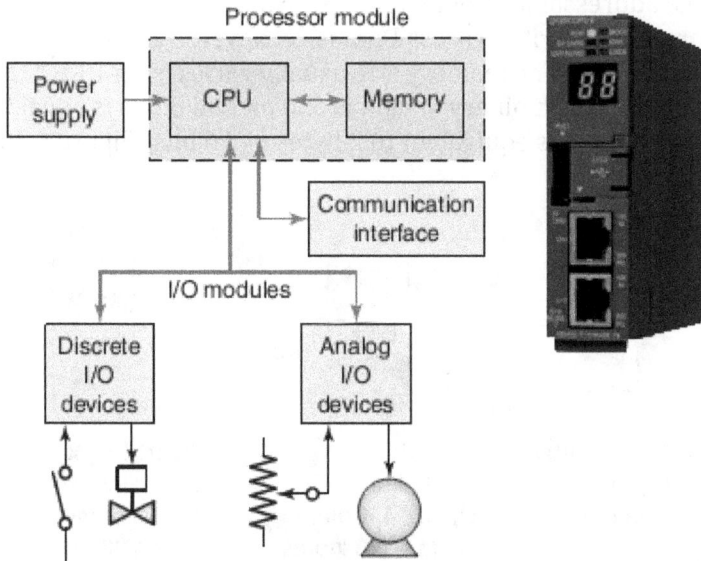

Figure 5.17 PLC CPU module segments.

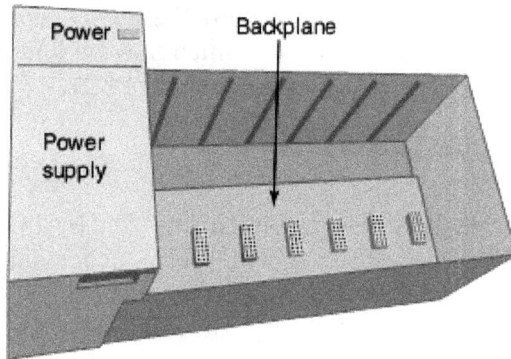

Figure 5.18 PLC power supply.

5.8 Designing memory

Memory is the element that stores information, program, and data in the PLC. PLC user memory includes space for user applications as well as addressable memory for storing data. Data is stored in memory under write process Data is retrieved from memory under reading process. The memory size usually varies from 1 KB to 32 MB (Figure 5.19). Each binary of data is one bit and 8 bits is equal to one byte, and a word is equivalent to 2 bytes or 16 bits (Figure 5.20).

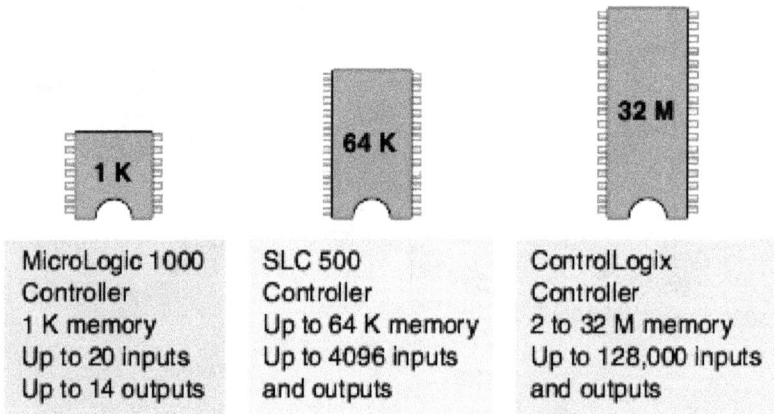

| MicroLogic 1000 Controller 1 K memory Up to 20 inputs Up to 14 outputs | SLC 500 Controller Up to 64 K memory Up to 4096 inputs and outputs | ControlLogix Controller 2 to 32 M memory Up to 128,000 inputs and outputs |

Figure 5.19 Different memory sizes.

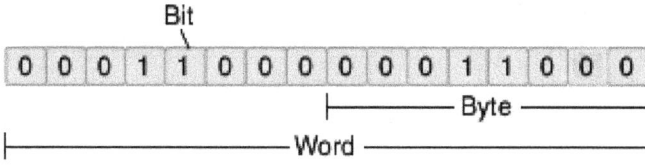

Figure 5.20 The concept of bits, bytes, and words.

5.9 Types of memory

1. ROM memory: stores programs and data that cannot be changed after memory is generated. ROM Used for PLC Operating System. The operating system is stored in the ROM memory by the PLC manufacturer and controls the software of the system through which the program is sent by the user to the PLC.

2. RAM Memory: This is called read/write memory (R/W) and is designed to write or read information in memory. RAM is used as a temporary data storage area that needs to be changed rapidly. The data stored in RAM is destroyed by power failure, so a backup battery is required to prevent data loss during power failure (Figure 5.21).

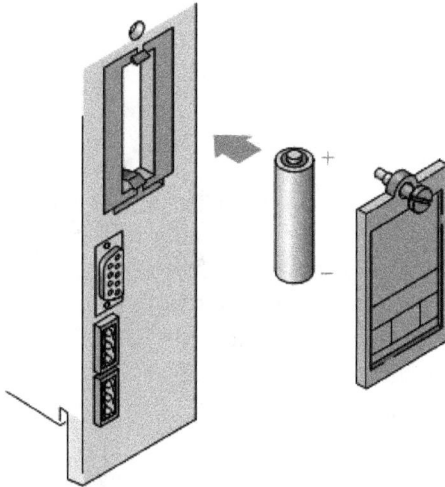

Figure 5.21 Backup battery for RAM.

3. EPROM Memory: This memory provides security for unwanted changes to the program. EPROMs are designed to read information stored in them but do not easily change without the use of special tools. For example, UV EPROMs are only cleaned by ultraviolet light. EPROM memory is used for backup, storage or transfer of PLC programs.

4. EEPROM Memory: Non-volatile memory having the same flexibility as RAM. The data is electrically written to it (instead of ultraviolet light). Requires no backup battery. EEPROM memory is typically used to store, backup, or transfer PLC programs (Figure 5.22).

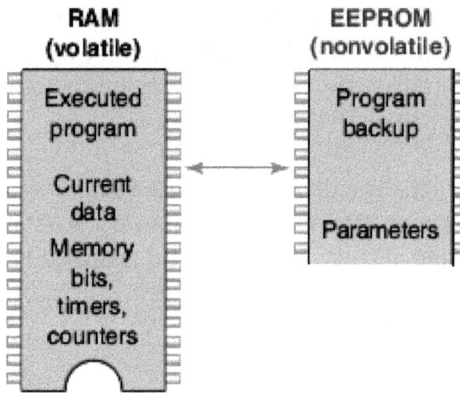

RAM
(volatile)

Executed
program

Current
data

Memory
bits,
timers,
counters

EEPROM
(nonvolatile)

Program
backup

Parameters

Figure 5.22 EEPROM memory for storing, backing up, or transferring PLC programs.

5. Flash EEPROM Memory: Like EEPROM, they are used for backup. Flash memories are very fast in storing and retrieving files and do not need to be physically separated from the processor for programming. Flash memories are sometimes built into the CPU module (Figure 5.23) which automatically backs up RAM. If the power supply is disconnected, the PLC will be restored without loss of data after the power is restored.

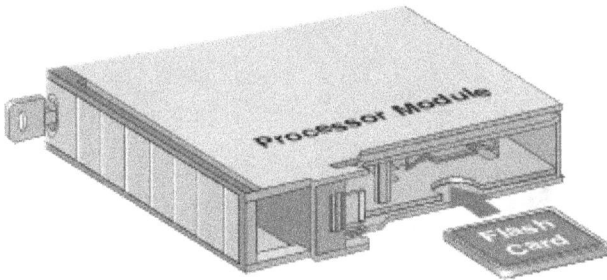

Figure 5.23 Flash memory installed in the socket on the processor.

5.10 PLC programming

A programming tool is needed to import, improve, and fix the PLC program. The simplest type of programming tool is shown in Figure 5.24. Compact handheld programmers are inexpensive and easy to use. It includes multifunction keys and an LCD or LED display.

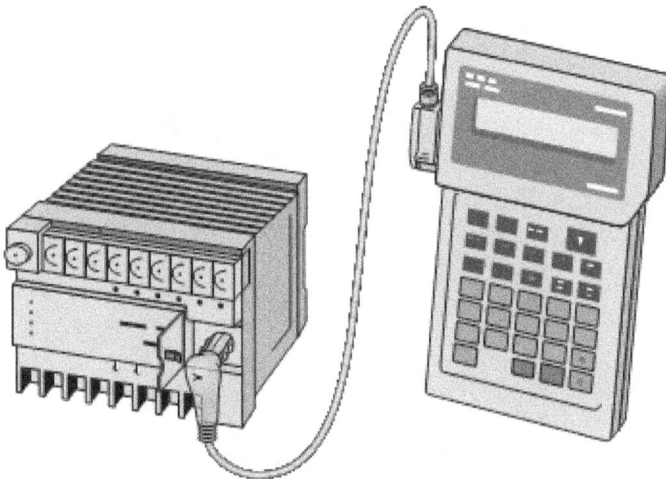

Figure 5.24 Manual Programming Tool.

The most common PLC programming methods are the use of a personal computer (PC) with the manufacturer's programming software (Figure 5.25). Examples of software capabilities include programming software, online and offline editing, monitoring, program documentation, PLC diagnostics, and control system troubleshooting.

Copies of reports produced in the software can be printed on a computer printer.

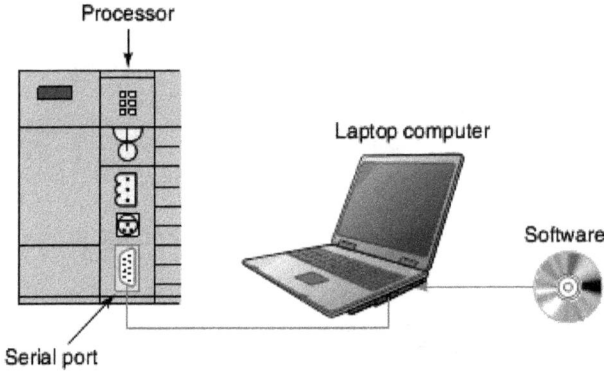

Figure 5.25 Personal computer used for PLC programming.

5.11 Data storage and retrieval

Printers are used to print CPU memory in the RLL format. The PLC has only one program in memory at a time. To change the PLC program, the new program needs to be imported directly from the keyboard or downloaded from the hard disk drive (Figure 5.26). Some CPUs support memory cartridges that provide EEPROM portable memory for the user program (Figure 5.27). The cartridge can be used to copy the program from one PLC to another PLC.

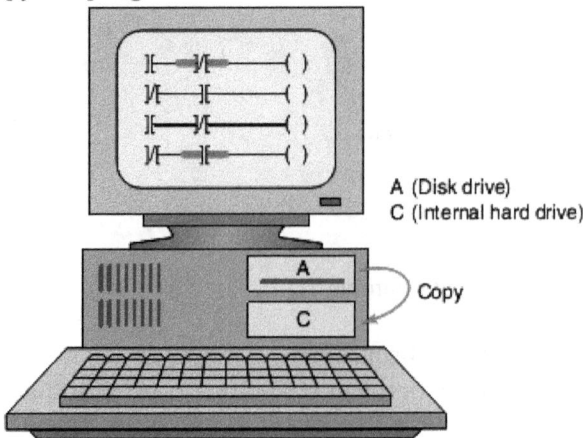

Figure 5.26 Copy Program to Hard Disk Drive.

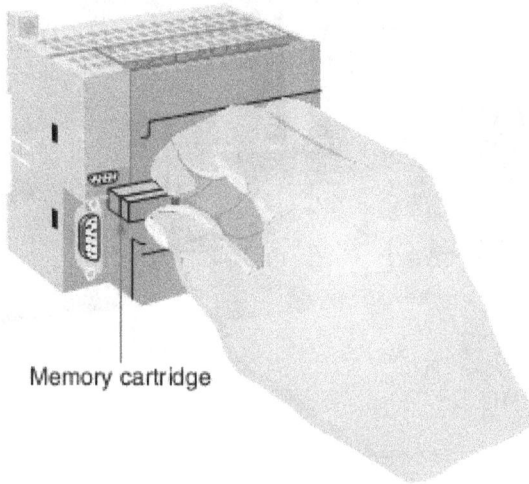

Memory cartridge

Figure 5.27 Memory cartridge.

5.12 HMIs

HMI (Human Machine Interfaces) can be used to communicate with the PLC and can be replaced with push buttons, selector switches, pilot lamps, thumbwheels and other control panel devices (Figure 5.28). The touchpad provides an operator interface similar to the traditional control panel. HMIs give operators and managers the ability to monitor real-time performance. Display PCs can be configured as follows:

• Replace the pads and pilot lamps with similar real icons. The operator must touch the screen to activate the push buttons.
• Display operations in graphical form for simplicity.
• Allow the operator to change the initial timer and counter counters with the graphical numeric keypad on the touch screen.
• Display alarms (indicating time and location)
• Show variables that changing over time.

Figure 5.28 HMIs.

5.13 Introduction to different kinds of industrial sensors

Introduction:

What is a sensor?

The sensor is a sensing element that converts the physical quantities such as pressure, humidity, temperature and so on to continuous (analog) or discrete (digital) quantities.

These sensors in a variety of measuring devices, robots, and analog and digital control systems such as PLC have made the sensors be a part of integral component of the automatic control system.

The sensors send various information about the status of the moving parts of the system to the control unit and change the performance of the devices.

From a control engineering perspective: The sensor is an integral component of feedback loop in closed loop systems.

From the point of view of instrumentation engineering: The first part of the measuring device that is directly related to the desired quantity, measures every moment of the resulting changes, and responds appropriately.

In general, the sensor is part of an integral component of an automatic control device that sends various information about the status of the moving parts of the system to the control unit.

Sensor Classification:

1- Thermal sensors:
Temperature sensors that are sensitive to temperature such as a thermometer.

2- Electric sensors:
Resistance Sensor (Ohmmeter) Current Sensor (Ammeter) Voltage Sensor (Voltmeter), Metal Detector and Radar Sensors.

3- Mechanical sensors:
Pressure Sensors (Altimeter, Barometer etc.)
Sensors for measuring the density, viscosity and fluidity of gases and liquids.
Mechanical sensors to detect position, acceleration, strength of objects.
Humidity sensors.

4- Chemical sensors:
Gas detection sensors such as oxygen, carbon monoxide and odor sensors.

5- Optical sensors:
Light sensors include semiconductors such as optical cells, photodiodes, infrared photo sensors, fiber optic sensors, etc.

6- Acoustic sensors:
Sensors used in microphones, headphones and used in the manufacture of robots.

7- Biological sensors:
Have a mechanism similar to mechanical sensors.
Many of these particular cells are sensitive to temperature, light, motion, magnetic fields, vibrations, pressure, and various physical quantities.
Biosensors, synthetic biosensors.

8- Sensors dependent on intra-atomic phenomena.

Overall sensor classification based on contact:

1- Contact sensors
2- Proximity or non-contact sensors

Contact sensors – micro switches

There are sensors that sense and act upon a piece of equipment. This is in a way that can absorb a relay, contactor or transmit an electrical signal to the input of a system. They are often used as a limit switch or home switch and as a safety guarantee for the system.

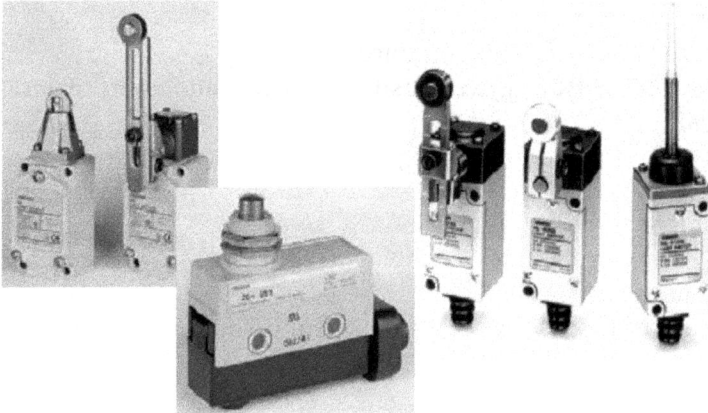

Non-contact sensors

Sensors are sensed and activated as one piece approaches it. This is such that it can attract a relay, contactor, or transmit an electrical signal to the input of a system.

Advantages of non-contact sensors:

1. High switching speed:
The sensors have a high switching speed compared to mechanical switches, such that some sensors work up to 25 kHz switching speed.

2. Long life span:
Due to no mechanical contact and lack of penetration of water, oil, dust and so on they have a long life.

3. No need for force or pressure:
Depending on the function of the sensor when approaching the piece, no force or pressure is required.

4. They can be used in different environments with hard working conditions:
These sensors are usable in high-pressure environments, high temperature, and acidic, oily, water and so on.
Due to the use of semiconductors, there is no bouncing noise in the output stage.

5. No noise during switching:
Due to the use of semiconductors in the output stage, no bouncing noise is created.

Some parameters in sensors:

1. Switching Frequency:
The maximum on/off number of a sensor in one second. (This unit is expressed in Hz).
This parameter is measured in accordance with DIN EN 50010.

2. Switching distance S:
It is the distance between the standard piece and the sensitive surface of the sensor during the switching operation.

Nominal switching distance Sn:
A distance that is defined in the normal state without considering variable parameters such as temperature, voltage etc.

Effective switching distance Sr:
It is the switching distance under rated voltage and temperature of
20C. Variable tolerances and parameters are also considered in this
case 0.9 <Sr <1.1 Sn.

Useful switching distance Su:
It is the switching distance in the range of permissible heat and voltage
0.81 Sn <Su <1.21 Sn.

Operational switching distance Sa:
It is the distance under which sensor performance is guaranteed
0 <Sa <0.81 Sn.

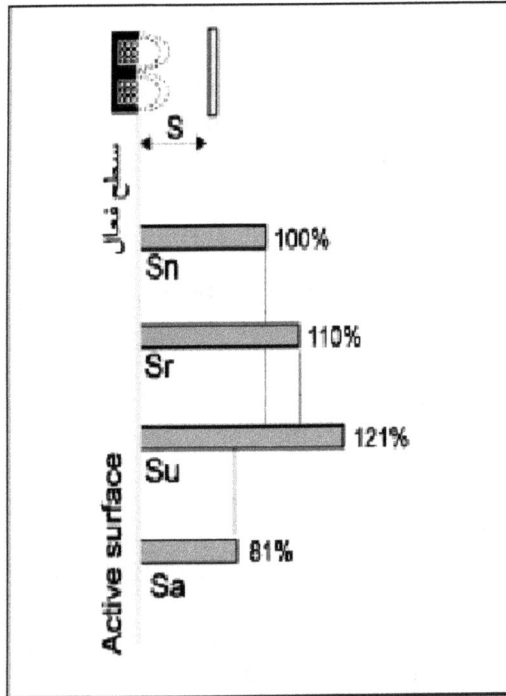

Inductive proximity switches:

Induction sensors are non-contact switches that act as switches
when approaching metal parts. Some of these sensors can output

voltage or analog current proportional to the distance from the sensor. These sensors are a good substitute for mechanical switches. Induction sensors have the following general characteristics:
• Produced in Normal Open (NO) and Normal Closed (NC) models.
• Special type of output (Decout) with PNP and NPN connectivity and NO and NC function (one sensor with capabilities of four sensors)
• All sensors are protected against overload and short circuit output.
• Produced in shielded and unshielded variants (with greater switching distance)
• Analogue type to measure work piece distance from sensor

The following table summarizes the inductive sensors and their operating distances.

Sensor Types & Operation Distance		φ3	φ4	M4	M5	D5	φ6.5	M8	M12	M18	M30	REC
Standard Type	shielded	0.6	0.8	0.6	0.8	0.8	1.5	1.5	2	5	10	
	unshielded						2.5	2.5	4	8	15	3
Non-standard Type	shielded	1	1.5		1.5		2	2	4	8	15	
	unshielded					1.5	4	4	8	12	20	6
Analog Type								0...4	0...6	0...10		
Operation Distance (mm)												

Optoelectronic Luminescence Sensors:

Luminas sensors are used to detect visible and invisible phosphorus signals using ultraviolet light.

These sensors are used to detect the luminas reference point even on perfectly reflective surfaces such as ceramics, metals, or mirrors. The following figure shows an example of the application of this sensor in the ceramic tile industry for the sorting lines. Among the tile lines, after being marked by phosphoric markers in terms of quality are classified by these sensors.

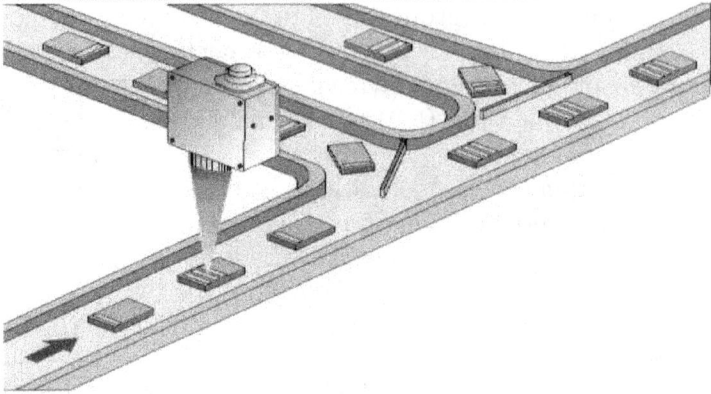

Optoelectronic Contrast sensors:

Contrast sensors (often known as color-signal sensors) are one-way sensors, but can detect two levels of varying degrees, for example, recognizing the presence or absence of front objects, and color on the wrapping paper. The light source can be reddish-green or white. If the light source is white, very low contrast can also be detected. Application of this sensor is in various industries such as packaging, petrochemical, printing, paint and other industries etc. The following figure shows one of the applications of these sensors.

Capacitive sensors are non-contact sensors that are capable of detecting metallic and non-metallic objects such as solids, liquids, powders.

The following figures show some of the applications of these sensors. The following are the characteristics of capacitive sensors.
• Metal body
• Protected output against short circuit and overload
• Type of NPN and PNP
• Cable and socket
• Adjust the sensitivity distance by a potentiometer 20 turns
• Specific type (self-monitoring)
• This sensor has output pulses that are used to control the sensor's sound. There may be cases where the sensor cable is disconnected or the sensor is unable to react due to a fault if there is a component in front of it. In the case of some pulses presence, it can detect sensor failure and prevents accidents from happening. The following figure shows some examples of these sensors.

Operation Distance
1.5...10 mm

Operation Distance
2...15 mm

Operation Distance
2...20 mm

Reflective optical sensors (One way):

In these sensors, which are the cheapest and simplest of sensors, the receiver and transmitter are in the same body.

The receiving object (without reflector use) returns the light sent from the transmitter, the transmitted light could be visible red - infrared or laser. There are also special types of these sensors with optical fiber.

However, because of these sensors, the received signal is weak, resulting in a decrease for Excess-Gain and will have a lower performance distance of up to two meters (the distance depends on the amount of reflection of the object).

Two-way Optoelectronic Through-beam Sensor:

These sensors are bi-directional on one side of the transmitter and on the other side of the receiver.

These sensors use laser-visible-infrared light. The transmitter transmits the generated waves and the receiver is installed in front of the transmitter. If there is no obstacle between the receiver and the transmitter, these waves will reach the receiver. However, if there is an obstacle, they will no longer reach the receiver. The following figure shows an application of these sensors.

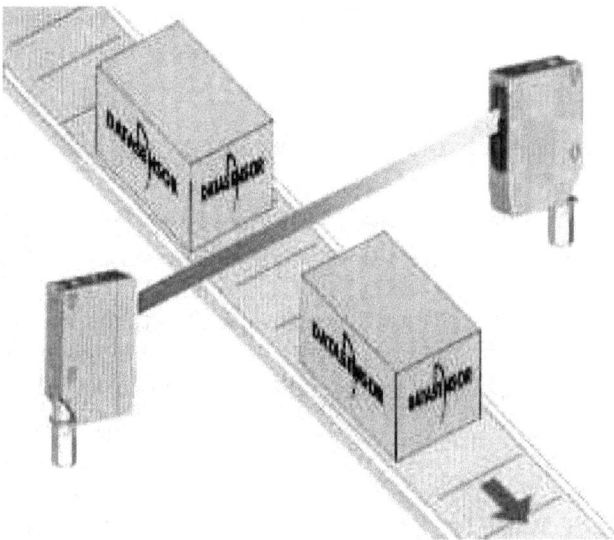

Laser sensors:

Laser sensors are manufactured in one-sided, two-sided, distance, reflective polarized, linear and one-way type with the background effect removing as shown below.

SL5	S50	S40	S60	S90	S80	Laser Class 1 Class 2
0...35 cm	0...35cm	4...15 cm	0...60 cm	0...60cm		One-sided Sensor
0...12m	0.1...16m		0.1...20m	0.1...20m		Reflective Polarized Sensor
0...60m	0...60m		0...60m	0...60m		Two-sided Sensor
				30...400 cm Analogue Class 2 Laser		Distance Sensor
		2...6 cm	5...10 cm	5...10 cm		One-way with BG Effect Removing

Area Sensors:

By creating a two-dimensional light surface, these sensors are capable of detecting and measuring the dimensions of objects. These sensors are built in DS1 and DS3 series. The DS1 series is easier and cheaper and the DS3 series more expensive and more capable.
Properties of surface scanners:

• Measuring the height of the components in absolute terms, using analog output
• Measuring relative height using analog output
• Transparent film detection and viewing of pieces of paper behind the transporter
• Ability to work in parallel with 6mm resolution
• Ability to work in the form of discontinuous rays

TX   RX

TX   RX

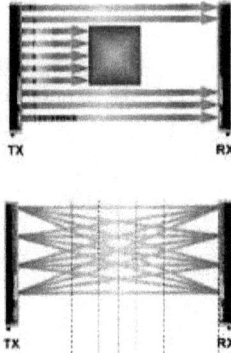

Color detection sensors:

Color recognition sensors in lines and assemblies are used to identify specific objects. The most important problem in these sensors is the detection of close or very bright colors.

For example, metallic used in the automobile industry makes color recognition difficult.

The number of colors it can detect, the ability to quickly change parameters, or to identify multiple colors simultaneously, determines the performance of the color detection sensor.

Conventional color detection sensors feature white LEDs with high light output that illuminate the modulated light on the target. Reflect light from the surface of the object to detect the main red, green and blue colors are analyzed and this information is used to determine if the parts are correct.

In some applications, the device user places a sample color in front of the sensor and schedules it to detect this color.

During the operation, the user can also define an interval for the color, thereby confirming the objects whose color is in this interval. This operation and adjustment of the upper and lower points is done by trial and error and does not have proper accuracy. Some color recognition sensors have only one output for confirmation and rejection. Thus, by connecting to the control system, they determine whether the object passing the front of the sensor is correct. However, in many other processes the need for deeper monitoring is more than just acceptance or rejection. The next generation of color detection sensors produce three more outputs that express the intensity of red, blue, and green. This capability makes the process more precise and smarter.

Applications:
New color recognition sensors can be used for applications that require expert color recognition systems. Automated systems can use the difference detected by color recognition sensors as input to the control system. For example, during the textile production process, color integration is very important. These sensors can monitor the density levels of all three primary colors continuously, and slight changes in color can be slowly corrected.

Ultrasonic sensors:

Ultrasonic sensors work by acoustic waves and reflections by the body.

They also use these sensors to measure distances. The reflected signal from the target goes back to the sensor and is measured, the time between sending and receiving the sound pulse is proportional to the distance between the sensor and the target. The performance of these sensors is independent of the color and polish of the surface of the object, and the ability to remove background suppression in these sensors is excellent. The outputs of these sensors are either analog (current or voltage) or digital. If the sensor head is impregnated with oil, adhesives, dust and so on. The sensor function will not be disturbed.

In case of segment deviation (up to 8 degrees), no change in measured value occurs.

Specifications of ultrasonic sensors:

• Digital type as simultaneous NPN and PNP
• Analog type (0-10) or (4-20)

- M18 body: Axial and Redial
- 10 - 30 VDC supply voltage
- Input: Socket cable
- Response time: four milliseconds
- Setting: by wire or the Teach-In button
- Measuring accuracy: 0.5 mm
- Normally opened or normal closed operation

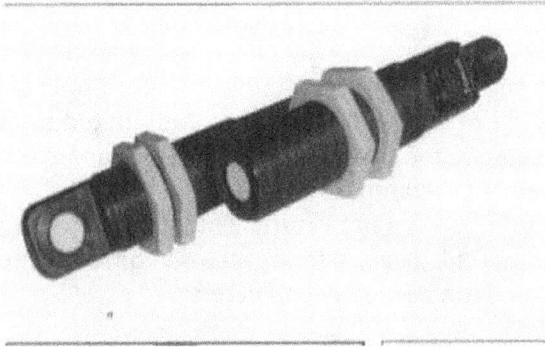

Temperature Sensors:

Several types of thermal sensors include:

1- Thermal Resistance Sensors
2- Thermoelectric sensors
3- PN connection heat sensors
4- Optical and acoustic heat sensors
5- Thermo mechanical sensors and actuators

Two types of heat resistance sensors are:

a) RTDs and
b) Thermistors (PTCs and NTCs)

Thermal Resistance Coefficient (RTDs):

Material	Resistivity ρ ($\Omega \cdot m$) at 20 °C	Conductivity σ (S/m) at 20 °C	Temperature coefficient (K^{-1})
Silver	1.59×10^{-8}	6.30×10^{7}	0.0038
Copper	1.68×10^{-8}	5.96×10^{7}	0.00404
Annealed copper	1.72×10^{-8}	5.80×10^{7}	0.00393
Gold	2.44×10^{-8}	4.11×10^{7}	0.00340
Aluminum	2.65×10^{-8}	3.77×10^{7}	0.0039
Calcium	3.36×10^{-8}	2.98×10^{7}	0.0041
Tungsten	5.60×10^{-8}	1.79×10^{7}	0.0045
Zinc	5.90×10^{-8}	1.69×10^{7}	0.0037

$$R = \frac{L}{\sigma S}$$

$$\sigma = \frac{\sigma_0}{1 + \alpha[T - T_0]}$$

$$R(T) = \frac{L}{\sigma_0 S}(1 + \alpha[T - T_0])$$

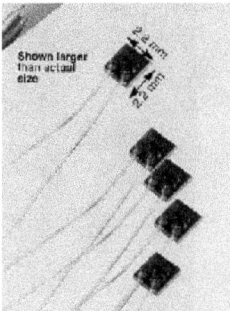

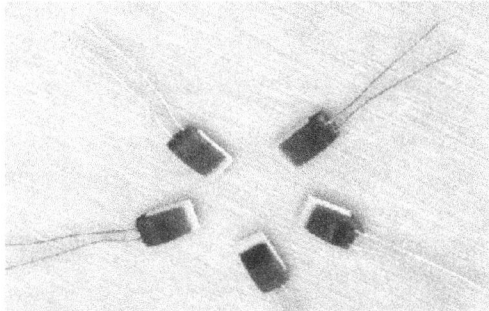

Thin Film RTD

Thermistors:

Most thermistors are NTC and sometimes PTC.

$$R(T) = \alpha e^{\beta/T}$$

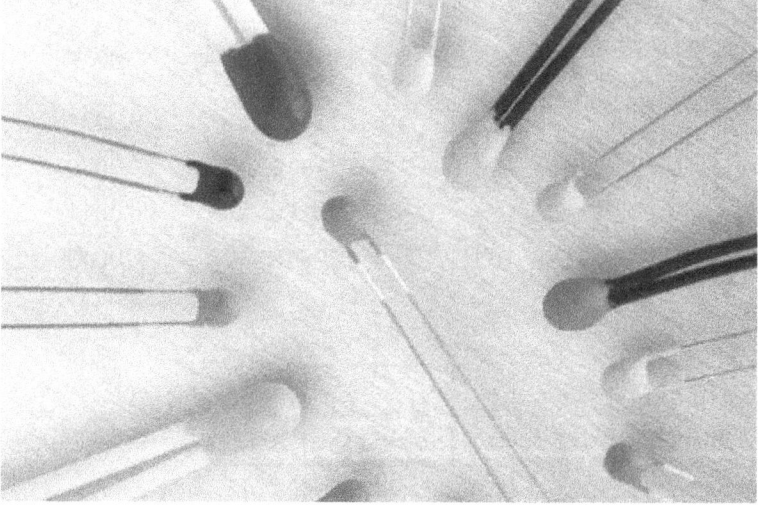

Epoxy Capsule Thermistors

Thermoelectric Sensors:

The most important thermoelectric sensor is thermocouple.

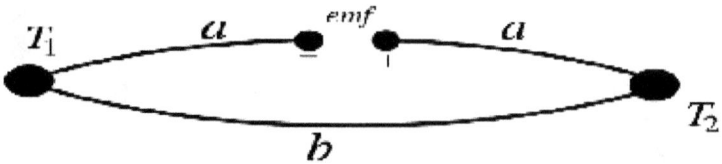

$$emf_A = \alpha_A(T_2 - T_1) \qquad\qquad emf_B = \alpha_B(T_2 - T_1)$$
$$emf_T = emf_A - emf_B = (\alpha_A - \alpha_B)(T_2 - T_1) = \alpha_{AB}(T_2 - T_1)$$
α ($\mu V/K$) is called Seebeck coefficient.

Type	Couples	Seebeck Coefficient
		µV/K
E	Chromel-Constantan	60
J	Iron-Constantan	51
T	Copper-Constantan	40
K	Chromel-Alumel	40
N	Nicrosil-Nisil	38
S	Pt (10% Rh)-Pt	11
B	Pt (30% Rh)-Pt (6% Rh)	8
R	Pt (13% Rh)-Pt	12

Seebeck Coefficients for Standard Thermocouples

The following are the most common types of thermocouples and the temperature range of each:

Code	Metals	Range	mV @ 100°C	Notes
S	PtRh/Pt	0–1400°C	0.645	Needs ceramic sheath
R	PtRh/Pt	0–1400°C	0.647	Needs ceramic sheath
J	Fe/CuNi	0–800°C	5.268	Attacked by oxygen or acids
K	NiCr/NiAl	0–1100°C	4.095	Avoid reducing agents
T	Cu/CuNi	−200°C to +400°C	4.277	Low temperature use
E	NiCr/CuNi	0–800°C	6.137	High output

PN Connection Heat Sensors:

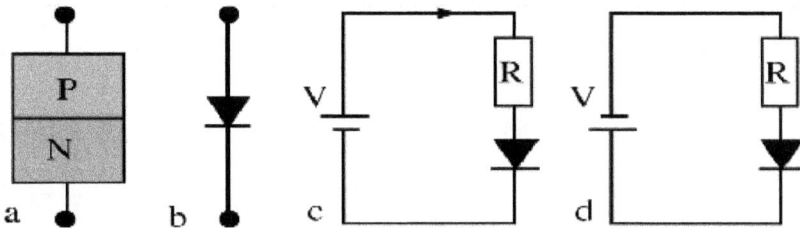

$$I = I_0 e^{qV/2kT}$$

$$V_f = \frac{E_g}{q} - \frac{2kT}{q} \, ln\!\left(\frac{C}{I}\right)$$

LM35 sensor

Optical Heat Sensors:

The spectrum of optical radiation is shown in the figure below:

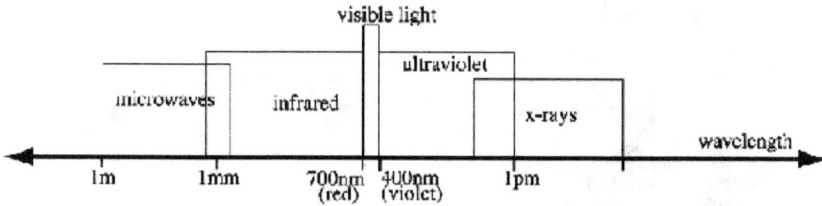

Photo-resistors:

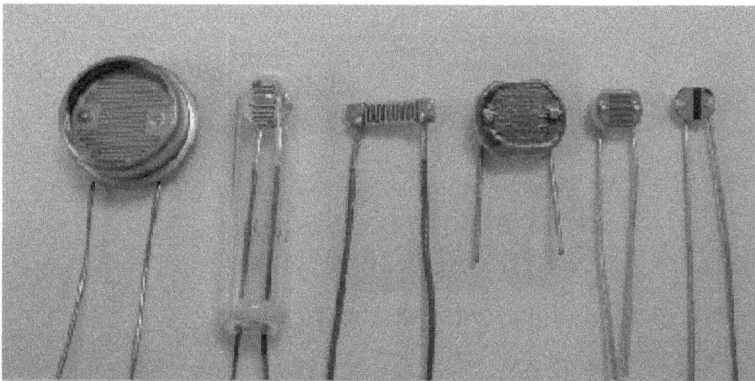

Various photoconductors (photo-resistors/photocells)

Photodiodes:

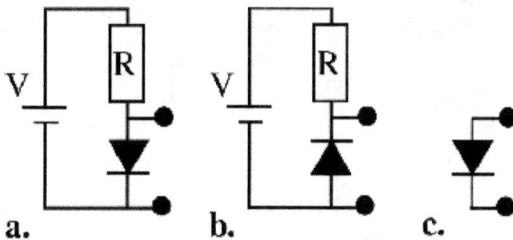

The photodiodes operate in two states:

a) Photoconductive state: The diode is in reverse bias and acts as a photoconductor.

b) Photovoltaic state: The diode does not bias and acts as a source (such as solar battery).

Photodiode array used in a scanner player

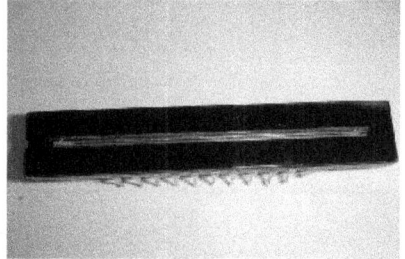

Photodiode array used in a CD player

Solar cells (photovoltaic)

Phototransistors:

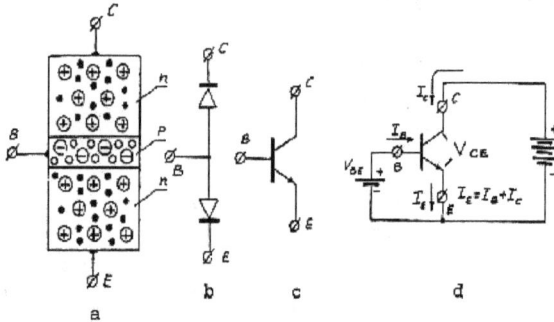

A typical phototransistor

Pyroelectric sensors:

Pyroelectric sensors consist of two parts: Passive Infrared (PIR) and Active Far Infrared (AFIR).

The example of a PIR sensor is shown below.

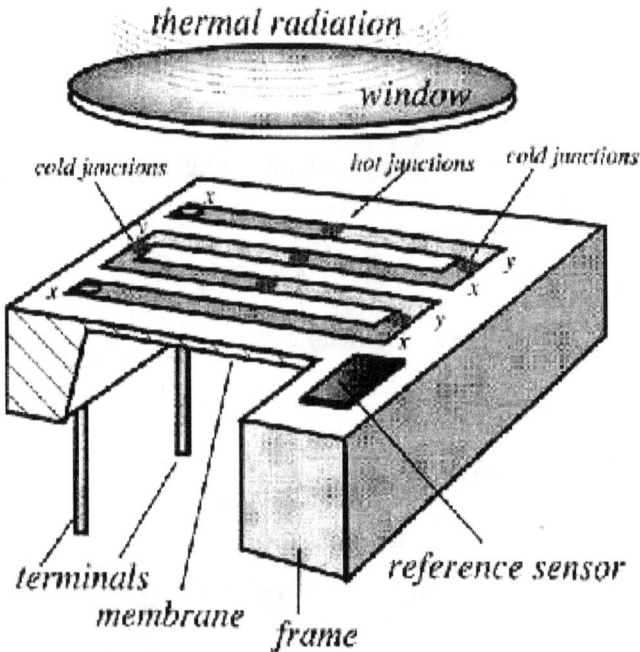

When a pyroelectric material is exposed to a ΔT temperature change, the ΔQ charge of its two ends is produced.

$$\Delta Q = P_Q A \Delta T$$

$$P_Q = \frac{dP_S}{dT}$$

The potential difference produced by the ΔV of the two sensors is obtained from the following relation:

$$\Delta V = P_V h \Delta T$$

And we have:

$$P_V = \frac{dE}{dT}$$

$$\frac{P_Q}{P_V} = \frac{dP_S}{dE} = \varepsilon_o \varepsilon_r$$

$$C = \frac{\Delta Q}{\Delta V} = \varepsilon_o \varepsilon_r \frac{A}{h}$$

$$\Delta V = P_Q \frac{\varepsilon_o \varepsilon_r}{h} \Delta T$$

At the above equations we have the following definitions:
A is the area of the sensor
P_Q is the pyroelectric charge coefficient
P_s is the spontaneous polarization of the material
h is the thickness of the crystal
P_V its pyroelectric voltage coefficient
E the electric field across the sensor
C is the sensor's capacitance

The table of pyroelectric materials is as follows:

Pyroelectric materials and some of their properties

Material	P_Q [C/m²/°K]	P_V [V/m/°K]	ε_r	Curie temperature [°C]
TGS (single crystal)	3.5×10^{-4}	1.3×10^{6}	30	49
LiTaO₃ (single crystal)	2.0×10^{-4}	0.5×10^{6}	45	618
BaTiO₃ (ceramic)	4.0×10^{-4}	0.05×10^{6}	1000	120
PZT (ceramic)	4.2×10^{-4}	0.03×10^{6}	1600	340
PVDF (polymer)	0.4×10^{-4}	0.4×10^{6}	12	205
PbTiO₃ (polycrystalline)	2.3×10^{-4}	0.13×10^{6}	200	470

TGS = TriGlycine Sulfate.

The structure of the piezoelectric sensor is as follows that can be connected in series or in parallel.

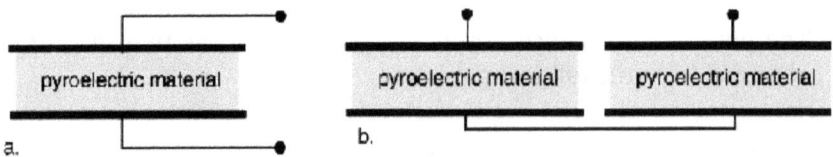

a.   b.

Another type of PIR sensor (motion sensor):

PIR motion detector

AFIR Performance Theory:

$$P = P_r + \phi$$

P is power supplied by an external source
P_L is power lost through conduction
Φ is radiation power sensed

$$P_L = \alpha_S [T_S - T_a]$$

$$T_m = \sqrt[4]{T_S^{\,4} - \left(\frac{1}{A\sigma\varepsilon}\right)\left[\frac{V^2}{R} - \alpha_S(T_S - T_a)\right]}$$

Power loss is:
α_s is a loss coefficient or thermal conductivity (which depends on materials and construction),
T_s the sensor's temperature
T_a the ambient temperature
Temperature of the radiating source is:
e is emissivity (total)
s is electric conductivity
T_a is ambient temperature
A is area of the sensor

Acoustic Heat Sensor:

The speed of sound is temperature dependent. The passage time of the heated material is measured. Most of the ultrasonic sensors are used for this purpose.

$$v_S = 331.5\sqrt{\frac{T}{273.15}} \qquad \left[\frac{m}{s}\right]$$

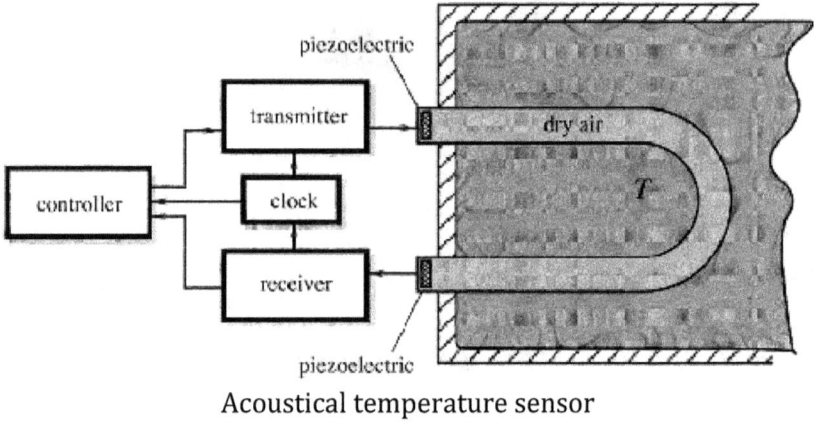

Acoustical temperature sensor

UT transmitter

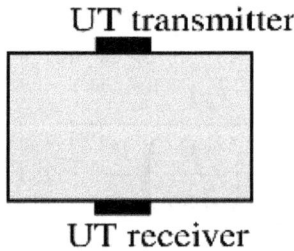

UT receiver

Acoustical temperature sensor

Thermo mechanical sensors and actuators:

1- Gas expansion temperature sensor:
 Heat causes gas expansion and presses the diaphragm on the sensor (strain gauge or potentiometer or a switch)

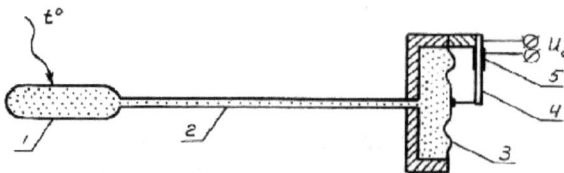

Gas expansion temperature sensor

2- Pneumatic sensor:

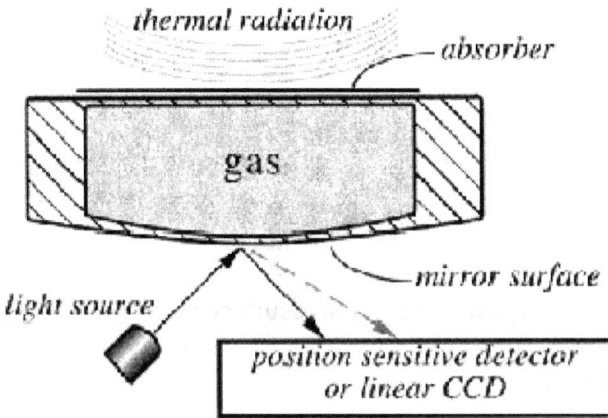

Thermo-pneumatic sensor

The gas expands in the flexible cell, which moves the mirror and the light reflects.

3- Thermal Expansion of Metals:

$$l_2 = l_1[1 + \alpha(T_2 - T_1)] \quad [\text{m}]$$

The linear expansion coefficients of the metals are as follows.

Substance	Coefficient of Expansion Per Degree Centigrade
Aluminum	25×10^{-6}
Brass or Bronze	19×10^{-6}
Brick	9×10^{-6}
Copper	17×10^{-6}
Glass (Plate)	9×10^{-6}
Glass (Pyrex)	3×10^{-6}
Ice	51×10^{-6}
Iron or Steel	11×10^{-6}
Lead	29×10^{-6}
Quartz	0.4×10^{-6}
Silver	19×10^{-6}

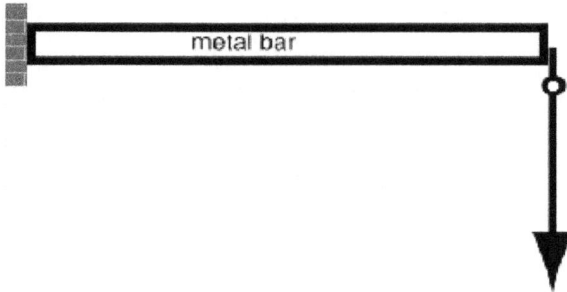

Direct dial indication

Instead of a dial, you can use a pressure sensor or a strain gauge.

4- Bimetallic sensors:

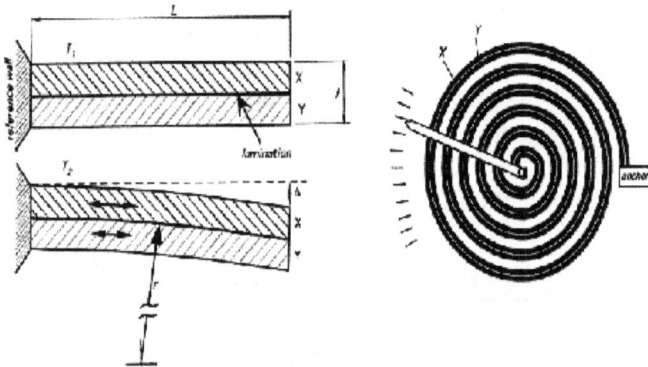

$$d = r\left[1 - \cos(\frac{180L}{\pi r})\right] \qquad [m]$$

$$r = \frac{2t}{3(\alpha_u - \alpha_l)T_2 - T_1}$$

d- Displacement for the bar bimetal
r- Radius of curvature
T_2- sensed temperature
T_1- reference temperature (horizontal position)
t- Thickness of bimetal bar

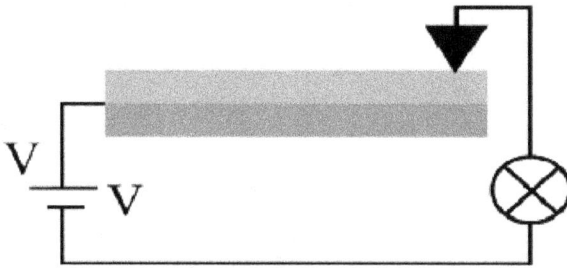

Bimetal switch (example)

Bimetal coil thermometer

Displacement sensors

1. Potentiometer
2. LVDT
3. Encoder

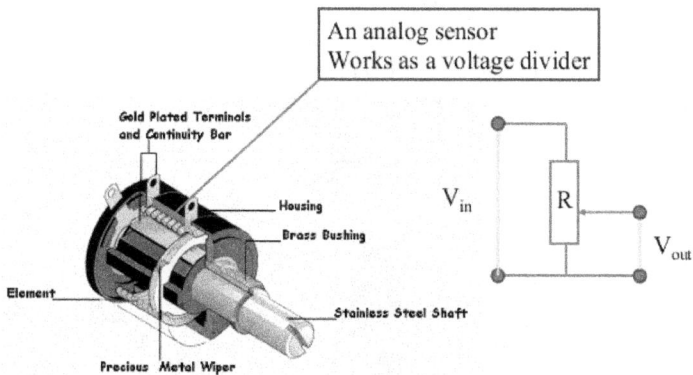

Potentiometer

Different types of potentiometers are wire wound, cermet and plastic film.

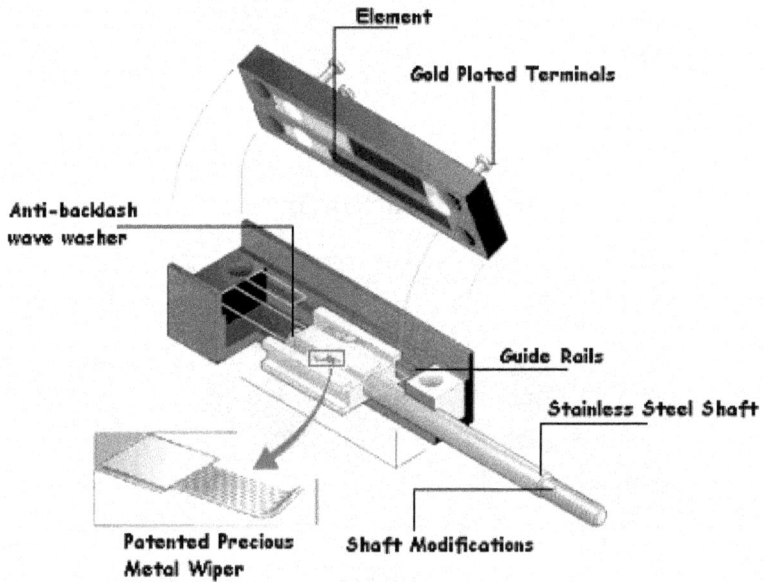

Linear Potentiometer

Linear Variable Differential Transformer (LVDT):

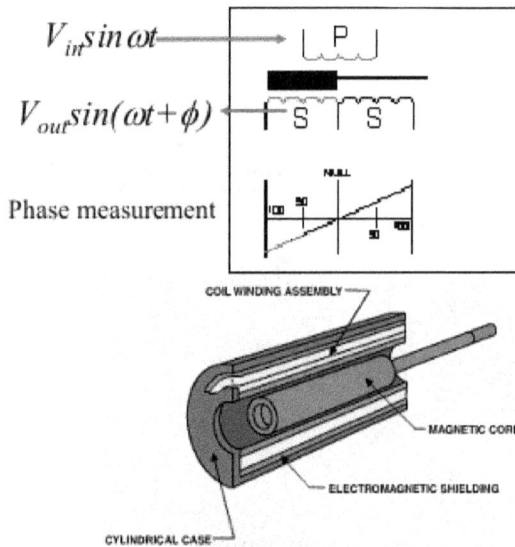

Signal Conditioning with LVDT:

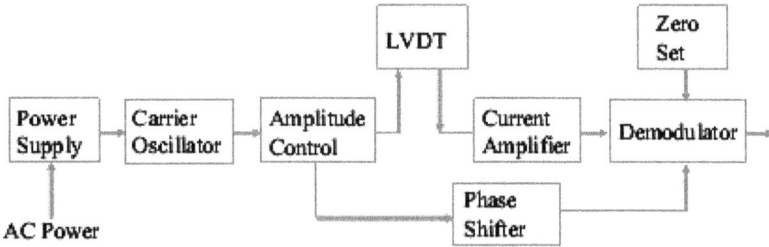

Examples of LVDT:

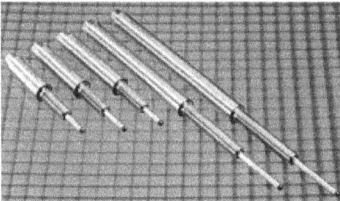

Free core LVDTs for
use in hostile environments
And total emersion

Spring-loaded
Standard for use
In hydraulic cylinders

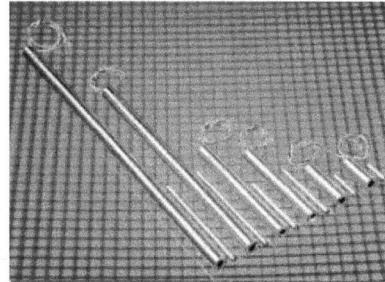

Encoders:

Optical encoders are digital sensors used to provide position feed-back to operators, including a plastic disk that is divided into light and dark sections, and pulses are generated as the disk rotates be-cause across two disks, there is an LED light source and a photo de-tector.

Encoder Signal Processing:

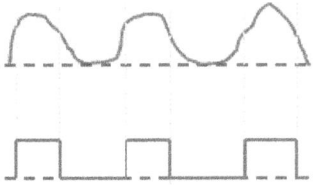

A square signal is required to digitally process the encoder signal.

Tachogenerator:

For measuring rotational speed a DC or AC tachogenerator is used.

Tacho generator for large
industrial plant (GE)

Force and pressure sensors:

Pressure and force are generally measured indirectly by surface deflection. The mechanisms used are:
1- Physical movement and measurement using an LVDT
2- Strain gauges (metal that change resistance under pressure)

3- Piezoelectric materials produce a current by deformation

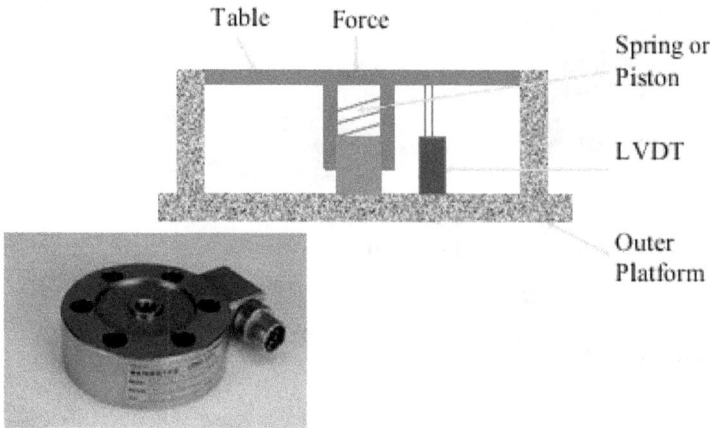

Table Force

Spring or Piston

LVDT

Outer Platform

LVDT Load Cell

Strain Gauge Bridge:

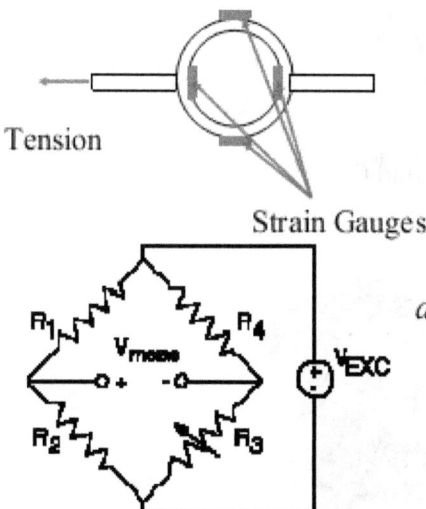

Tension

Strain Gauges

$$GF = \frac{\Delta R/R}{\Delta L/L} = \frac{\Delta R/R}{\varepsilon}$$

$$\Delta R = R \cdot GF \cdot \varepsilon$$

$$\frac{V_{meas}}{V_{exc}} = \left(\frac{R_3}{R_3 + R_4} - \frac{R_2}{R_1 + R_2} \right)$$

$$assume\ R_1 = R_2, R_4 = R_G,$$

$$R_3 = R_G + \Delta R$$

$$then\ \varepsilon = \frac{-4V_{meas}}{GF \left(2V_{meas} - V_{exc} \right)}$$

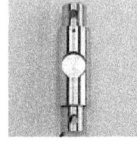

Examples of load cells:

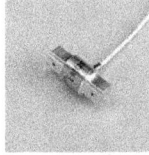

Subminature   Reaction torque   Axial load cell
Load cells   load cell

Piezoelectric Cells:

Piezo Load Cells

Pressure Sensors:

The Bourdon Tube

Diaphragm pressure sensor

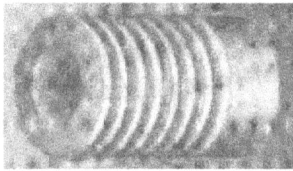

Bellow pressure sensor

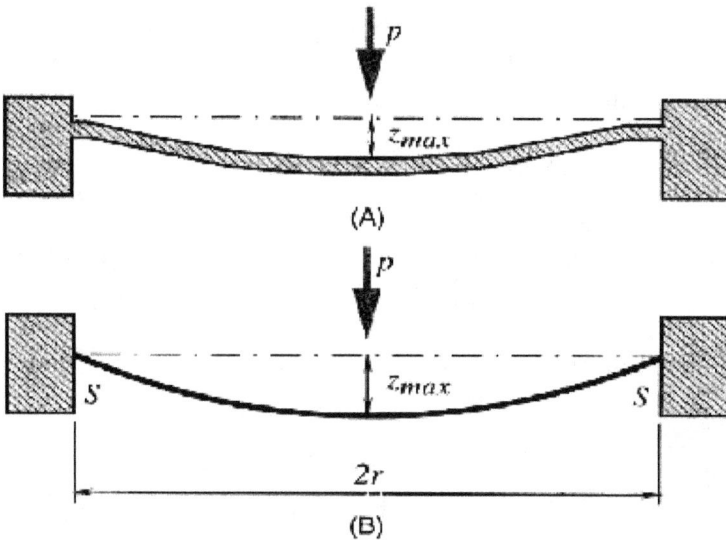

Membrane and thin plate

$$d_m = \frac{r^2 P}{4S} \qquad\qquad \sigma_m = \frac{S}{t}$$

d_m is the radial tension
S is the radial tension area
P is the applied pressure difference between the top and bottom of the membrane
r is its radius
t is its thickness

If the thickness t cannot be ignored:

$$d_m = \frac{3(1-v^2)r^4 P}{16Et^2} \qquad\qquad \sigma_m = \frac{3r^2 P}{4t^2}$$

E is the Young's modulus
V is the Poisson's ratio

Piezoresistive pressure sensors:

Pizoristor is a semiconductor strain gauge. Most pressure sensors use it instead of a conductive strain gauge. Metallic or resistive strain gauges are only used at higher temperatures or in special cases.

Structure of piezoresistive pressure sensors:

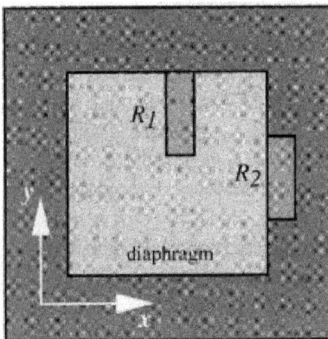

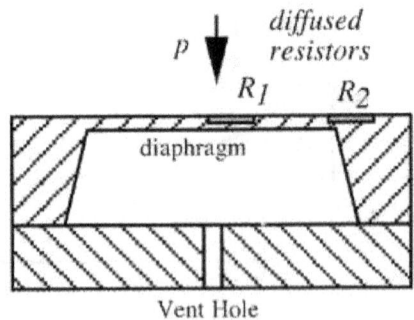

$$\frac{\Delta R_1}{R_1} = -\frac{\Delta R_2}{R_2} = \frac{1}{2}\pi(\sigma_y - \sigma_x)$$

π is an average sensitivity (gauge) coefficient and σ_x and σ_y are the stresses in the transverse directions.

Note: In pressure sensors, deformation is measured in the same way as force sensors:

a) Spring (manometer)
b) Piezo distortion
c) Strain gauges

Acceleration Sensors:

Acceleration is also measured by the force applied by an accelerated mass.
a) Distortion of a piezo
b) Movement of a cantilever stand
c) Drawing on a restricted mass
d) They are mainly used to measure vibration.

Types of flow meters:

The percentage of industrial flow meters is as follows:

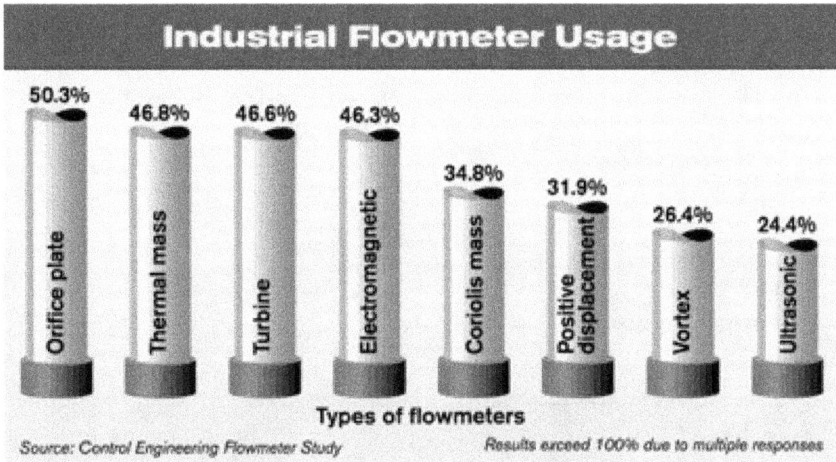

Industrial Flowmeter Usage

- Orifice plate: 50.3%
- Thermal mass: 46.8%
- Turbine: 46.6%
- Electromagnetic: 46.3%
- Coriolis mass: 34.8%
- Positive displacement: 31.9%
- Vortex: 26.4%
- Ultrasonic: 24.4%

Types of flowmeters

Source: Control Engineering Flowmeter Study   Results exceed 100% due to multiple responses

The types of fills are as follows:

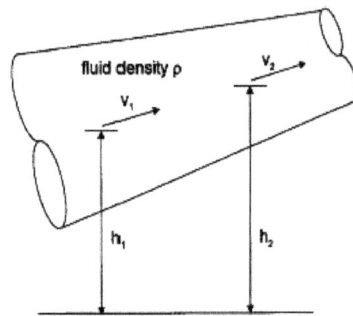

$$\frac{P_1}{\rho g} + \frac{v_1^{\,2}}{2g} + h_1 = \frac{P_2}{\rho g} + \frac{v_2^{\,2}}{2g} + h_2$$

1- Orifice Plate

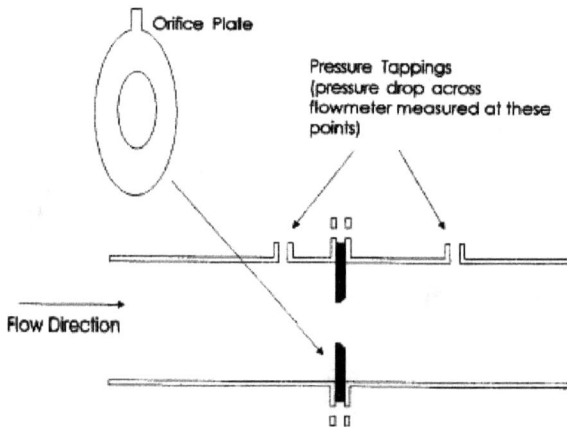

2- Venturi tube:

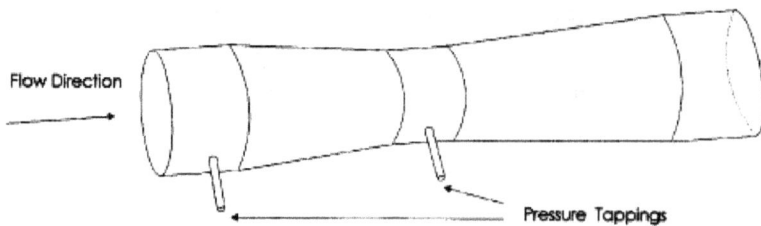

3- Flow Nozzle

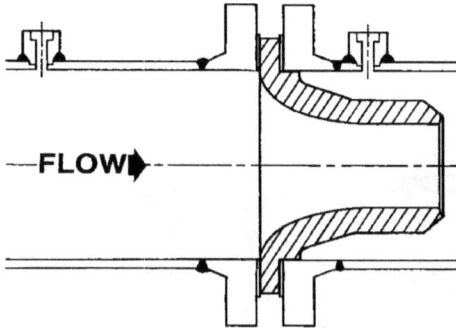

This flow meter is based on pressure difference, which is used at higher speeds and pressures, and it is better to use for gases than liquids.

4- Wedge Meter

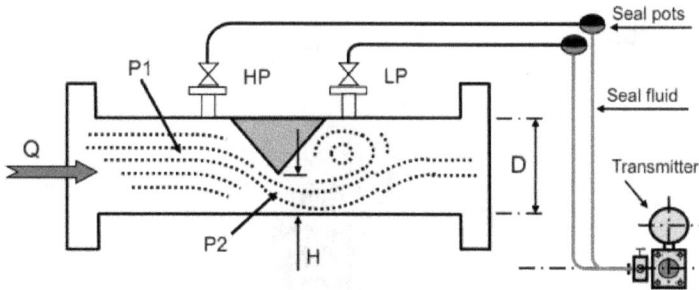

Like orifice plates, they are used for any type of gas or liquid. They are generally used in high viscosity dirty fluids with solid suspended particles.

5- V-Cone

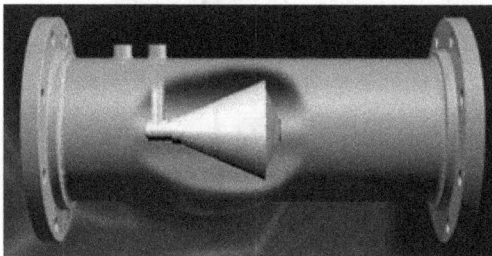

It is similar to pressure difference flow meters. These flow meters are used in high viscosity fluids.

6- Pitot Tube

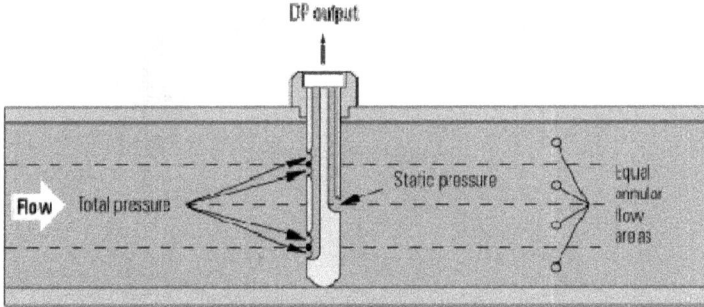

These flow meters use dynamic pressure difference to calculate the flow. Orifice plates record all static and dynamic pressures. The difference in pressure between the two causes the dynamic pressure in the tube to be used to calculate the flow.

7- Rotameter

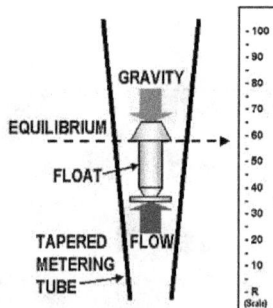

$$Q = kA\sqrt{gh}$$

where:
Q = volumetric flow rate, e.g., gallons per minute
k = a constant
A = annular area between the float and the tube wall
g = force of gravity
h = pressure drop (head) across the float

With h in the rotameter constant, A is a direct function of the flow velocity Q. Thus, the rotameter designer can determine the cone of the pipe in such a way that the height of the float in the tube is equal to the flow rate.

8- Thermal Mass

$Q = WCp$ (T_2-T_1) and therefore
$W = Q/Cp$ (T_2-T_1)
Q = Heat Transfer
W = Mass Flow Rate
Cp = Specific Heat of Fluid
T_1 = Temperature Upstream
T_2 = Temperature Downstream

A) Immersion Heater

B) Externally-Heated Tube

Thermal Mass Flowmeter Designs

A) Bypass Uses Small Percent of Stream

B) Temperature Profile

The Bypass Flowmeter Design

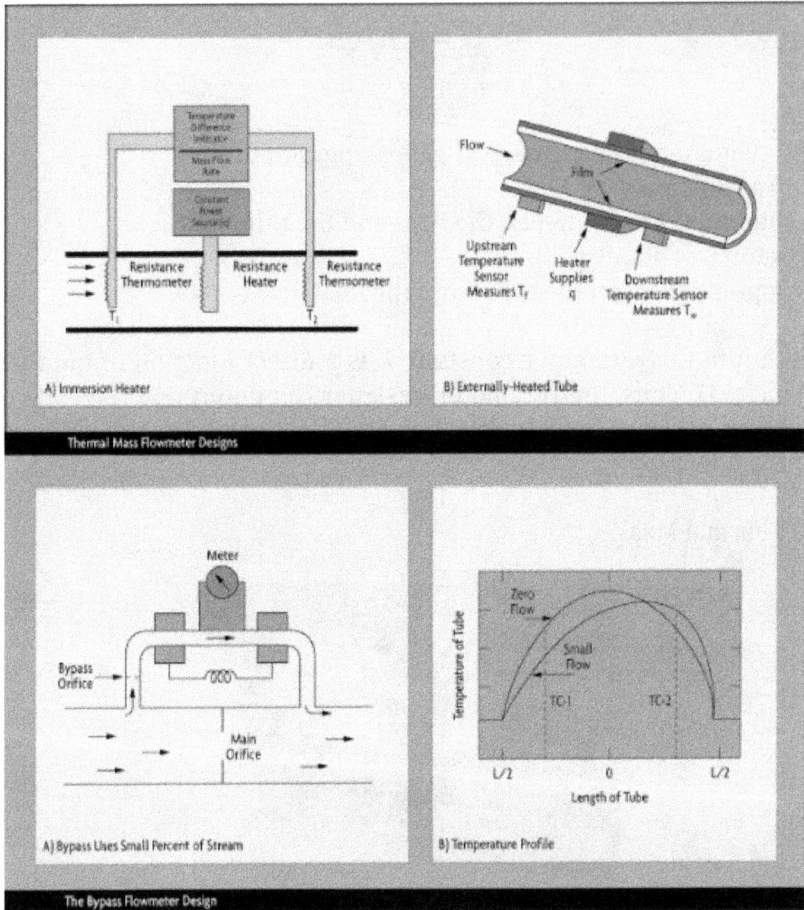

9- Turbine flow meter

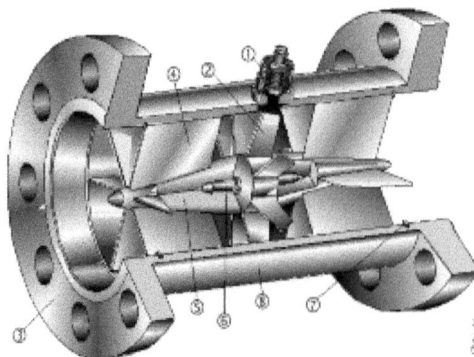

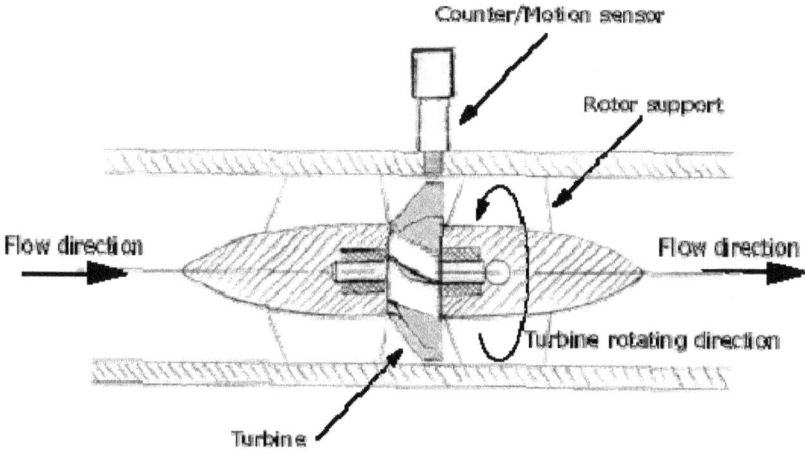

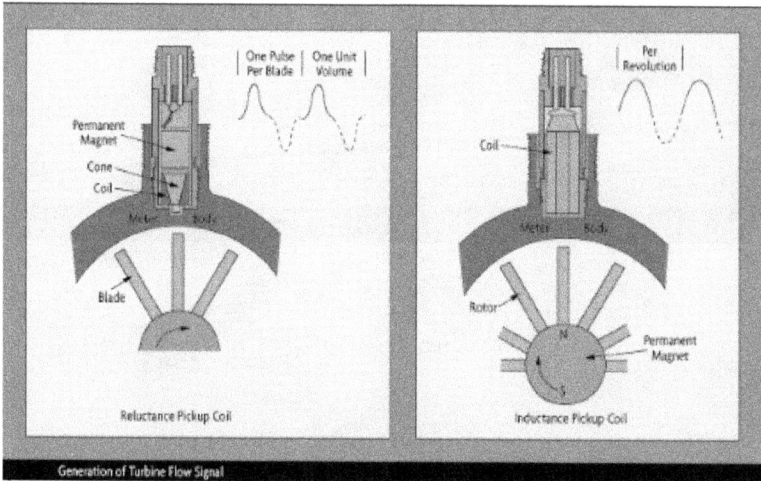

10- Electromagnetic flow meter

E = BDV/C
E = Induced Voltage
B = Magnetic Field Strength
D = Inner Diameter of Pipe
V = Average Velocity
C = Constant

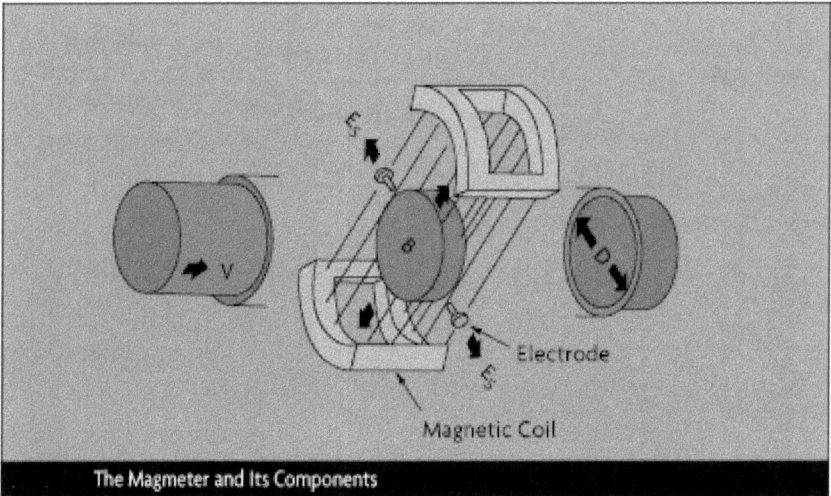

The Magmeter and Its Components

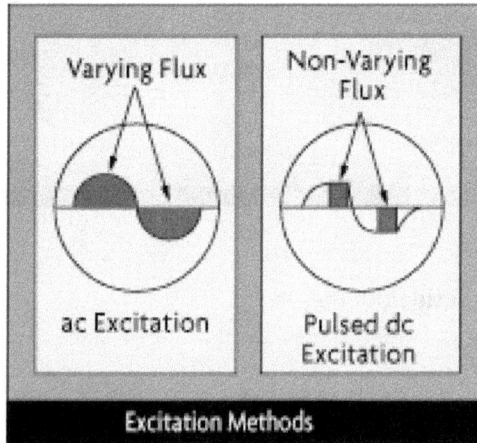

Excitation Methods

11- Corriolis flow meter

Newton's 2nd Law of angular motion states that
$\gamma = I\alpha$ and defines that $H = I\omega$ and since by definition $I = mr^2$

Then $\gamma = mr^2\alpha$ and then $H = mr^2\omega$

Since $\alpha = \omega/t$ then becomes $\gamma = mr^2 \omega/t$ and solving mass flow rate, m/t we get

$m/t = \gamma/r^2\omega$ also divide $H = mr^2\omega$ by t then $H/t = m/t (r^2\omega)$

H = Angular Momentum

I = Moment of Inertia

ω = Angular Velocity

Y = Torque

α = Angular Acceleration

r = Torque of Gyration

m = Mass, t = Time

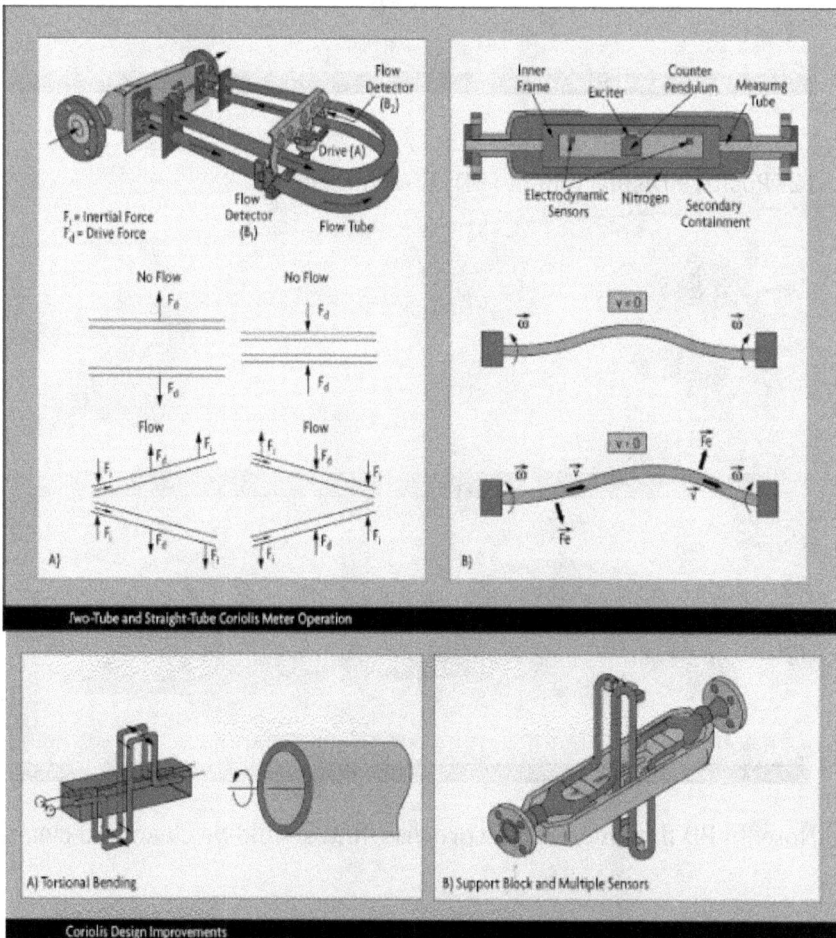

Two-Tube and Straight-Tube Coriolis Meter Operation

A) Torsional Bending

B) Support Block and Multiple Sensors

Coriolis Design Improvements

Installation Variations of the Coriolis Meter

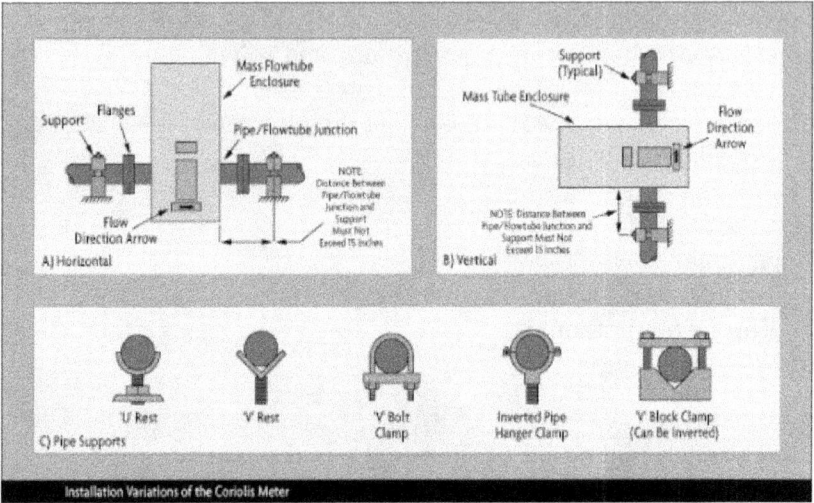

12- Positive Displacement (PD) flow meter

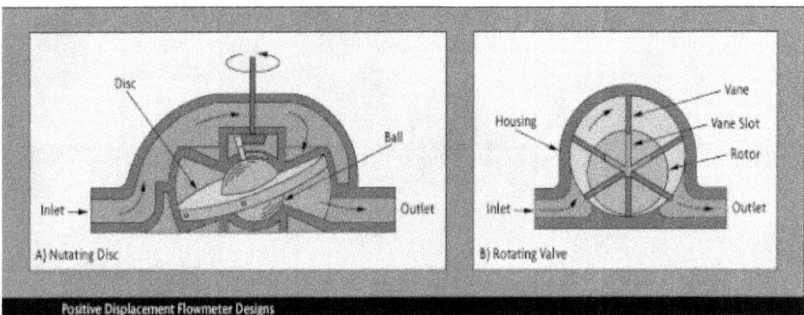

Positive Displacement Flowmeter Designs

Note: In PD flow meters, the process fluid should be clean and clean.

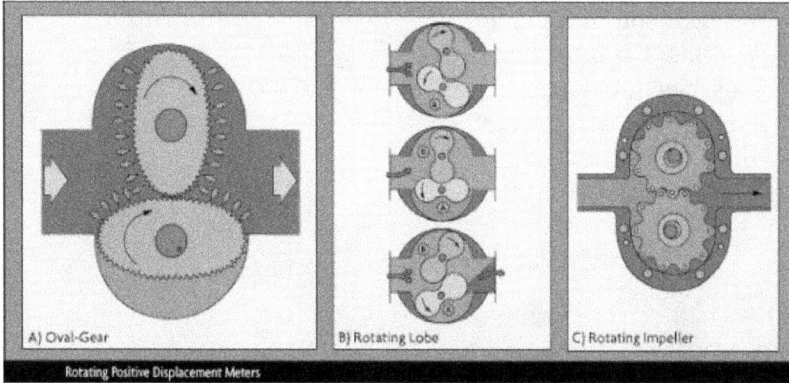

A) Oval-Gear

B) Rotating Lobe

C) Rotating Impeller

Rotating Positive Displacement Meters

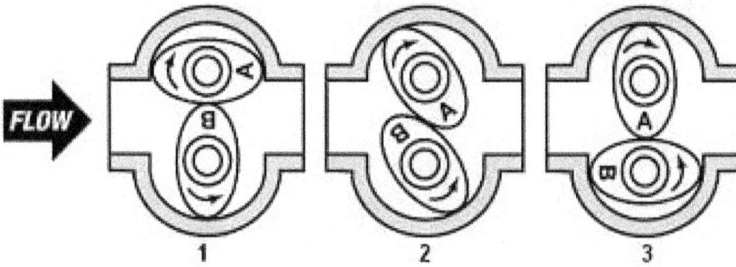

FLOW

1   2   3

The flow is proportional to the gear speed.

13- Vortex flow meter

The relationship between vortex frequency and fluid velocity is:
St = f (d/v)

Q = AV = (AfdB)/St
Q = fK
St = Strouhal Number

f = Vortex Shedding Frequency
d = Width of the Bluff Body
A = Cross Sectional Area
V = Average Fluid Velocity
B = Blockage Factor
K = Meter Coefficient

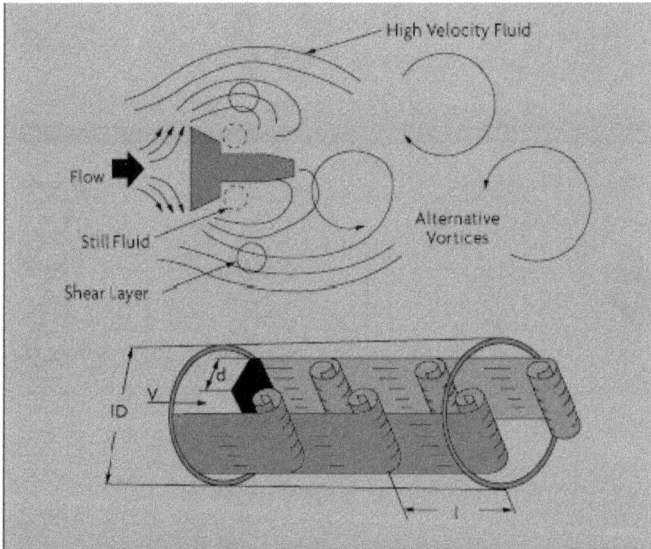

Vortex Meter Calculation of Flow Velocity

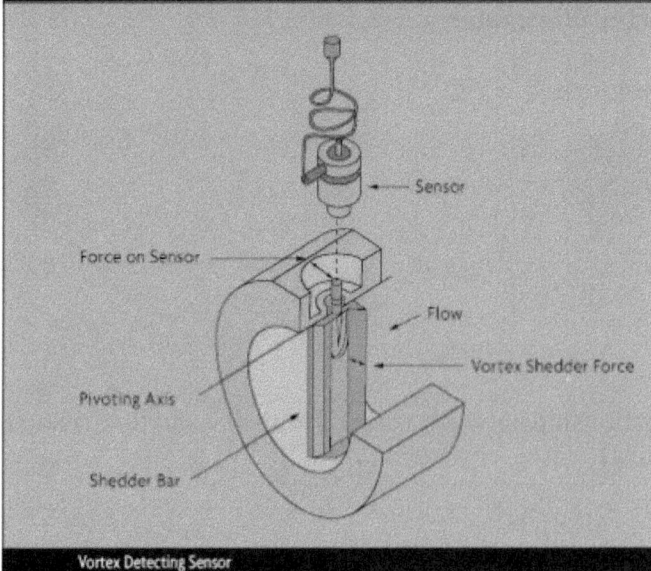

Vortex Detecting Sensor

14- Ultrasonic flow meter

Doppler Effect Ultrasonic flow meter
Transit Time Difference Ultrasonic flow meter

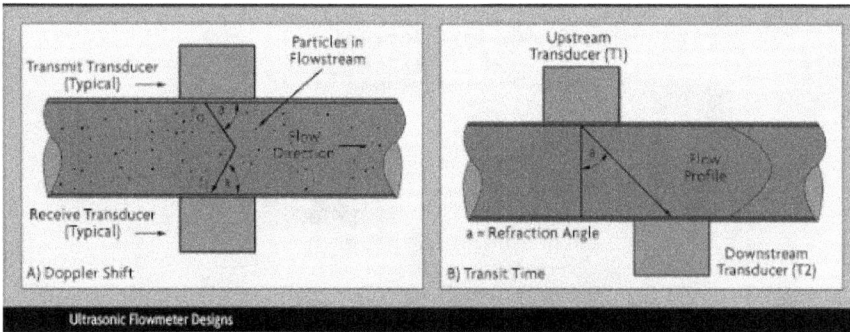

Ultrasonic Flowmeter Designs

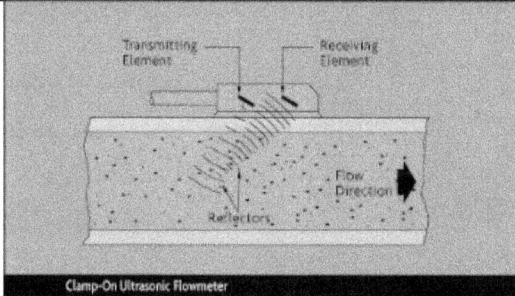

Clamp-On Ultrasonic Flowmeter

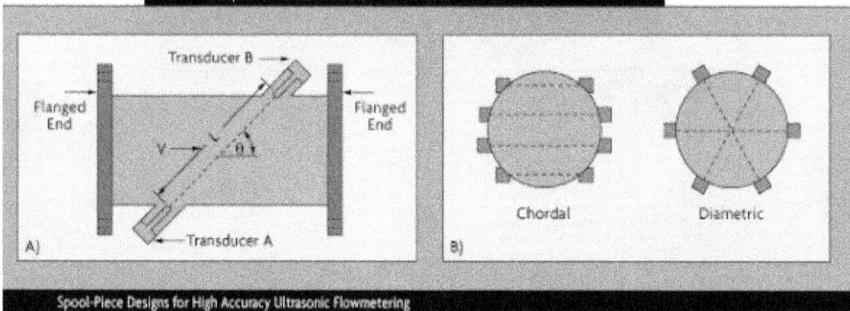

Spool-Piece Designs for High Accuracy Ultrasonic Flowmetering

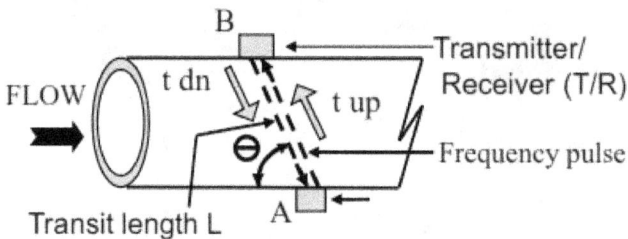

The average velocity, Vm, is calculated as follows:

$$V_m = \frac{L}{2}\cos(\theta)\left[(T_{AB} - T_{BA})/(T_{AB} \cdot T_{BA})\right]$$

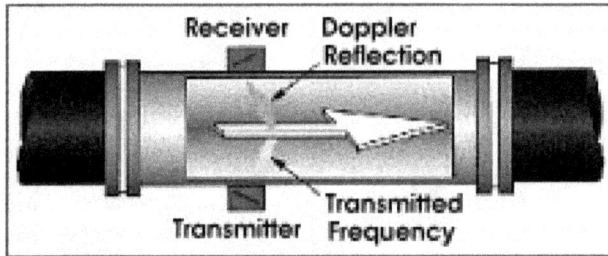

$$\Delta f = 2 f_T \sin(\theta)\frac{V_F}{V_S} \qquad V_F = \frac{\Delta f}{f_T}\frac{V_T}{\sin(\theta_T)} = K\Delta f$$

$$\frac{\sin(\theta_T)}{V_T} = \frac{\sin(\theta)}{V_S}$$

$$Q = KV_F D^2$$

D= inside diameter of the pipe

V_T = sonic velocity of transmitter material
θ_T = angle of transmitter beam
K = calibration factor
V_F = flow velocity
Δf = Doppler frequency shift
V_S = sonic velocity of fluid
f_T = transmitter frequency
θ = angle of f_T entry into liquid

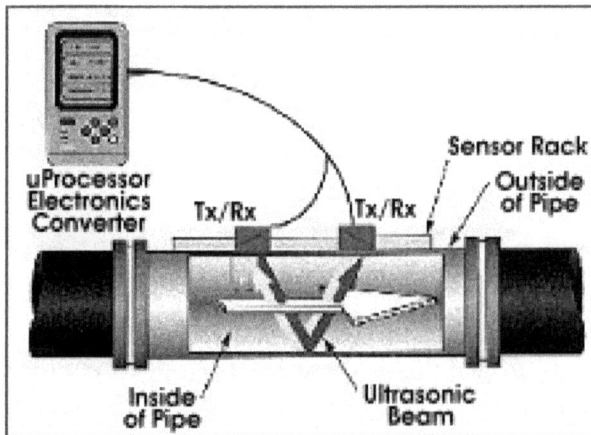

$$V = \frac{KD}{\sin(2\theta)}\frac{1}{(T_0 - \tau)^2}\Delta T$$

V   = mean velocity of flowing fluid

K   = constant

D   = i.d. of pipe

θ = incident angle of ultrasonic burst waves

T_0 = zero flow transit time

$\Delta T = T_2 - T_1$

T_1 = transit time of burst waves from upstream transmitter to downstream receiver (blue path in Figure 3)

T_2 = transit time of burst waves from downstream transmitter to upstream receiver (yellow path in Figure 3)

τ = transit time of burst waves through pipewall and lining.

Comparison and Selection of Flow Meters:

		Ultrasonic	Electromagnetic	Differential Pressure	Vortex
Measuring Media	Fluid	✓	✓	✓	✓
	Gas	X	X	✓	✓
	Vapor	X	X	✓	✓
	Slurry	X	O	X	X
Application	Control	✓	✓	✓	✓
	Monitor	✓	✓	✓	✓
	Supply	X	✓	X	X
Operating Condition	Temperature	-40 to 200 °C	-20 to 120 °C	-40 to 600 °C	-10 to 200 °C
	Pressure	-	-1 to 2MPa	-0.1 to 42MPa	5MPa
	Pressure Loss	None	None	Yes	Yes
	Rangeability	Large	Large	Large	Large
Installation Condition	Bore ∅	13 ~ 6,000mm	2.5 ~ 300mm	25 ~ 3,000mm	4 ~ 100mm
	Upstream/Downstream	10D/5D	5D/2D	10D/5D	7D/3D
	Piping Works	Not Required	Required	Required	Required
	Explosion Proof	✓	X	✓	X
Performance	Accuracy	± 0.5 % of Rate	± 0.5 % of Rate	± 2.0 % of FS	± 1.0 ~ 3.0 % of Rate
	Velocity Range	-32 to 32m/s	0 to 15m/s	-	0.3 to 4m/s

Chapter 6

Working with analog signals and HMI in the SIMATIC S7 TIA Portal

6.1 Introduction

PLCs must be able to work with continuous or analog signals. Typical analog signals are 0-10 VDC or 4-20 mA. Analog signals are used to display variable values such as speed, temperature, weight and level. The PLC cannot process these signals in analog form. The PLC must convert the analog signal into a digital signal. The PLC analog modules convert standard voltage and current analog values to 12-bit digital, for example. Digital values are sent to the PLC to be stored as word. In addition, analog modules are used with thermocouple and RTD sensors to achieve a high level of accuracy in temperature measurement. For example, in the figure below, a hand is attached to the load cell. A load cell is a device that receives the weighted variable values and converts it to the values of the output voltage or current variable. In this example, the load cell converts a weight value to a 0-10 VDC output. Load cell outputs are 0-10 VDC for inputs 0-500 Lbs. The 0-10 VDC load cell output is connected to the PLC analog module input.

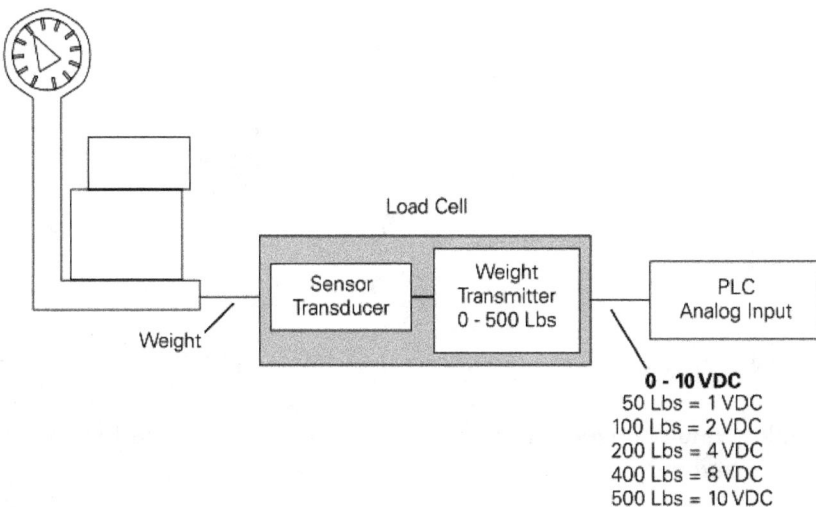

Applying this example can be in a single-gate conveyor system that deflects packets with variable weights. As the packages move along the conveyor, their weight is measured. A pack whose weight is greater than a certain value is guided along a conveyor route. A packet weighing less than a certain amount is guided to another conveyor route that can be inspected later for lack of content.

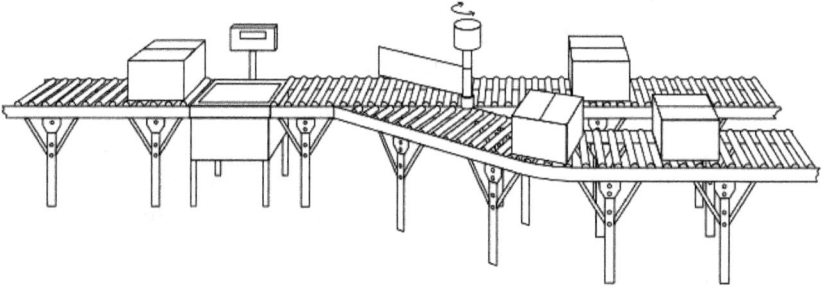

Analog outputs are used in applications requiring control of field devices that respond to constant voltage or current levels. Analog outputs are used as a variable reference for valves control, chart registers, electric motor drives, analog meters and pressure transducers. Like analog inputs, a transducer via a control generally controls analog outputs. The transducer receives the voltage signal and, depending on the need, amplifies, attenuates it or changes it to another signal to control the device. In the figure below, a 0-10 VDC signal controls the 0-500 Lbs analog gauge.

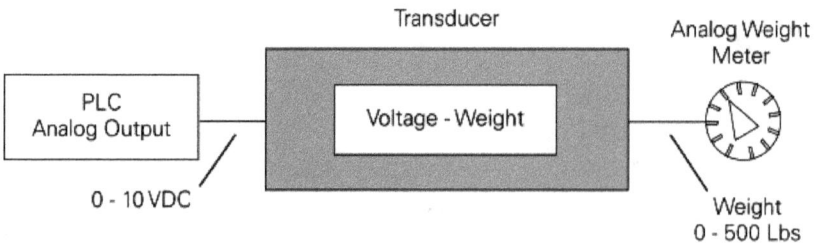

6.2 Working with analog signals in the SIMATIC S7 TIA Portal

Analog value conversion for PLC processing is the same for analog inputs and outputs. The range of digital values is as follows.

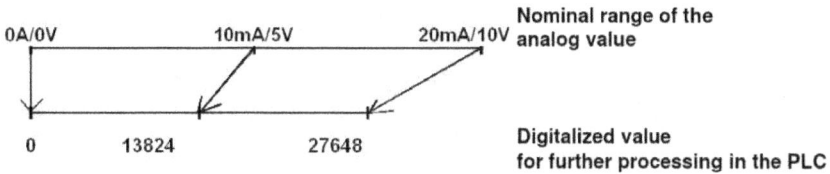

These digital values are usually normalized for further processing in the PLC. Consider, for example, the level of a tank.

- A sensor measures the tank surface and converts it to a 0-10 VDC voltage signal. 0 V is proportional to the level of 100 liters and 10 V is proportional to the level of 1000 liters.

- The sensor is connected to the SIMATIC S7-300 analog input. The signal is input to the FC1 function and normalized.

- It is planned to monitor the maximum acceptable level of 990 liters and the minimum acceptable level of 110 liters.

Analog input address and digital output are as follows.

Address	Symbol	Data Type	Comment
%IW 64	AI_level_tank1	Int	Analog input level Tank1
%Q 0.0	Tank1_max	Bool	Indication > 990 liters
%Q 0.1	Tank1_min	Bool	Indication level < 110 liters

The following steps explain how to implement the above example as a project in the SIMATIC S7-300.

1- Double click (→Totally Integrated Automation Portal V11)

2- In the portal view (→ Create new project → Tank_Analog → Create)

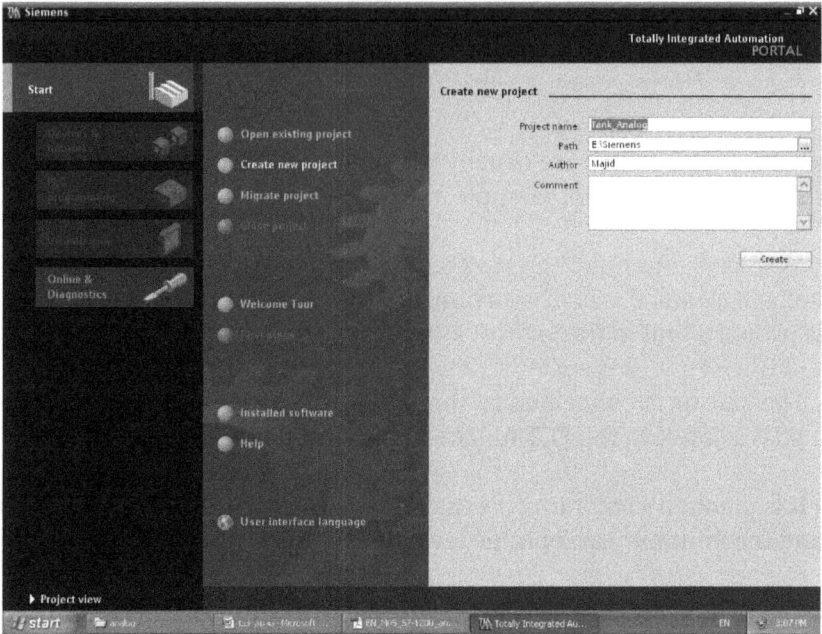

3- Configure a device (→ First Steps → Configure a device)

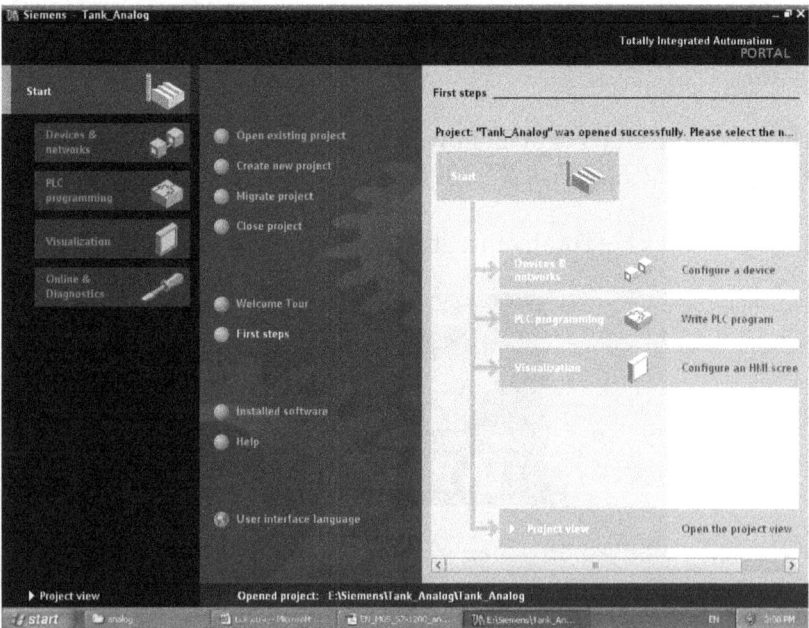

4- Add new device (→ Add new device → Control_tank → CPU317-2 PN/DP → 6ES7 .....→ Add)

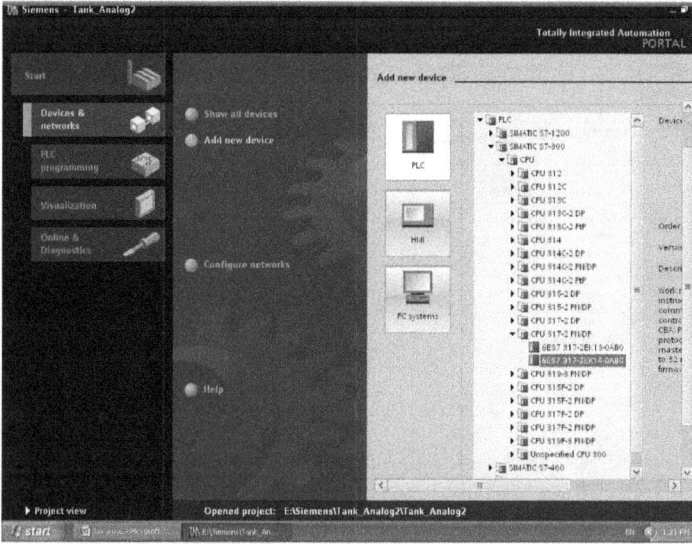

5- With drag & drop we add from the catalog the signal board (→ Catalog → Signal board → AO1 x 12Bit → 6ES7 332-...), (→ Catalog → Signal board → AI1 x 12Bit → 6ES7 331-...), (→ Catalog → Signal board → DI8/DO8×24VDC→ 6ES7 323-...)

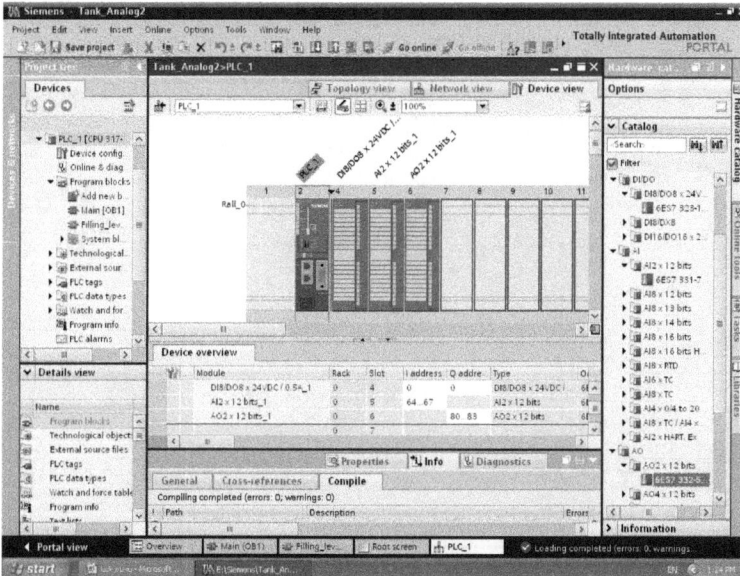

6- IP address and the subnet mask have to be set (→ Properties → General → PROFINET interface → Ethernet addresses →IP address:192.168.0.1 → Subnet mask: 255.255.255.0)

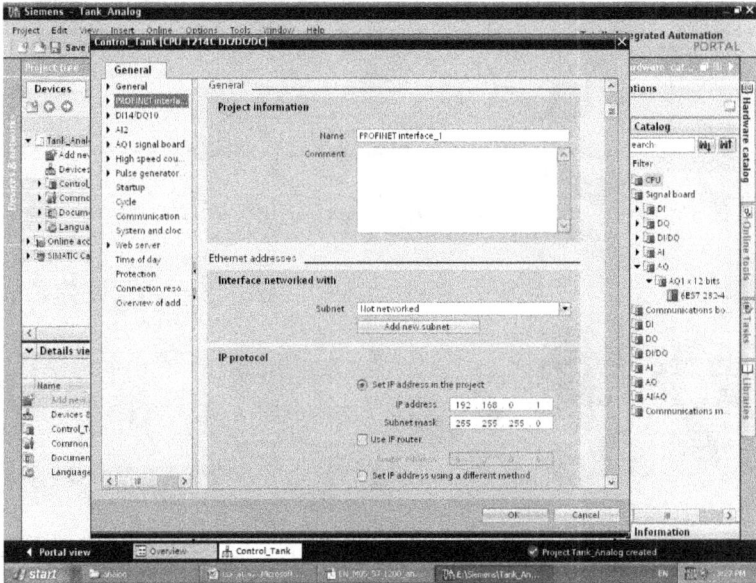

7- PLC tags (→ Control_tank[CPU317-2 PN/DP] → PLC tags→ PLC tags)

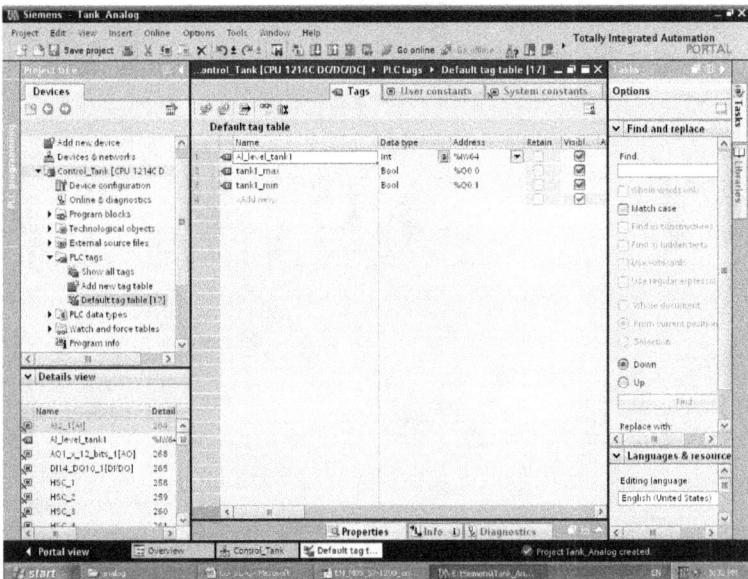

8- Add new block (→Control_tank[CPU317-2 PN/DP] → Program blocks → Add new block), (→ Function (FC1) → Filling_level_tank1 → FBD → OK)

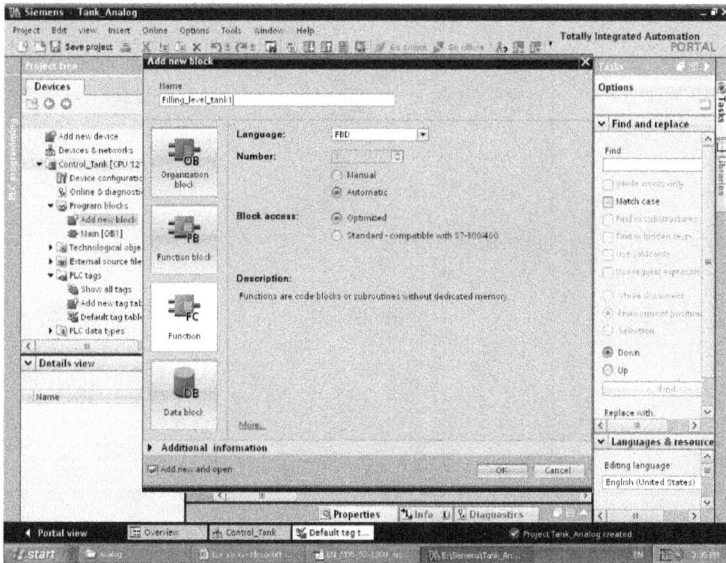

9- Block parameters that are the interface of the block for calls in the program.

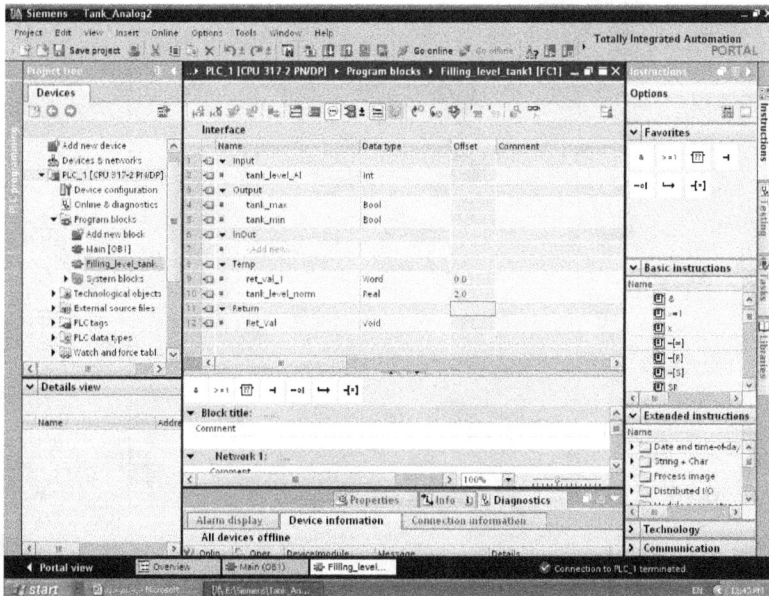

10- Enter the program by using the variable names (variables are identified with the symbol '#'.)

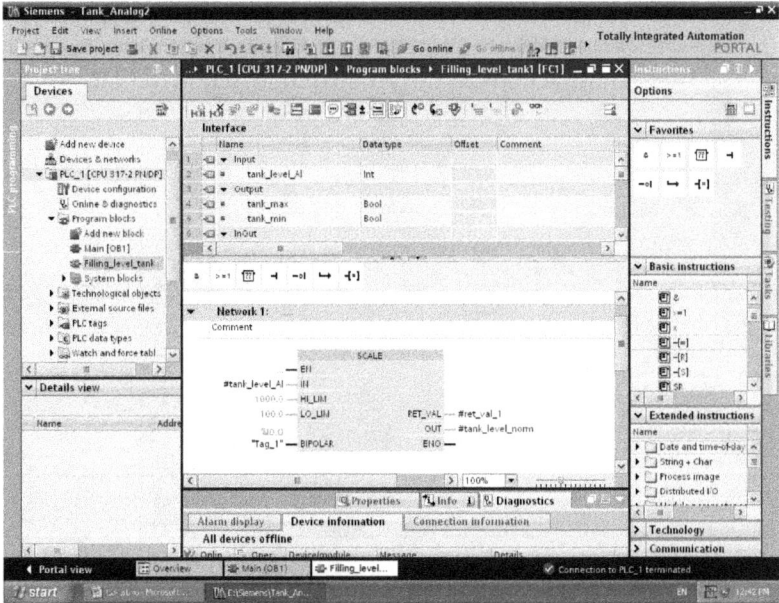

11- Continue entering the program.

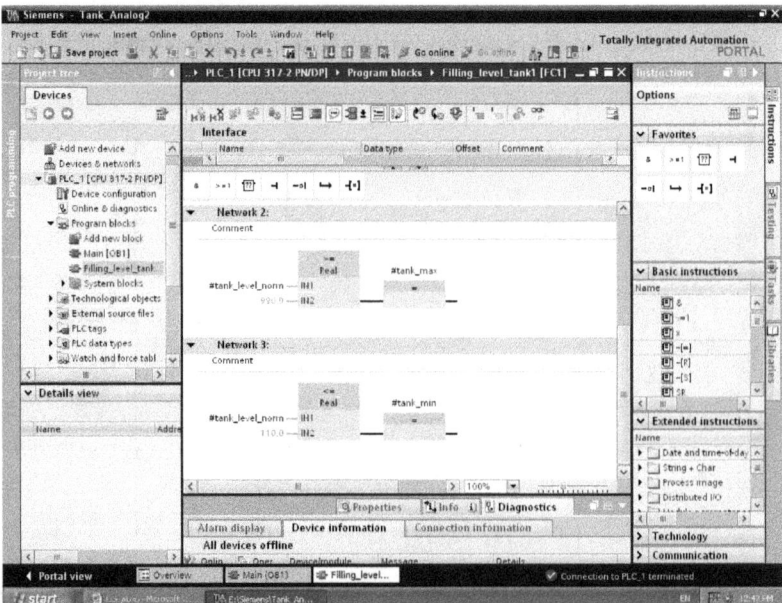

12- Right click Main[OB1] (→ Properties), Language (→ FBD → OK)

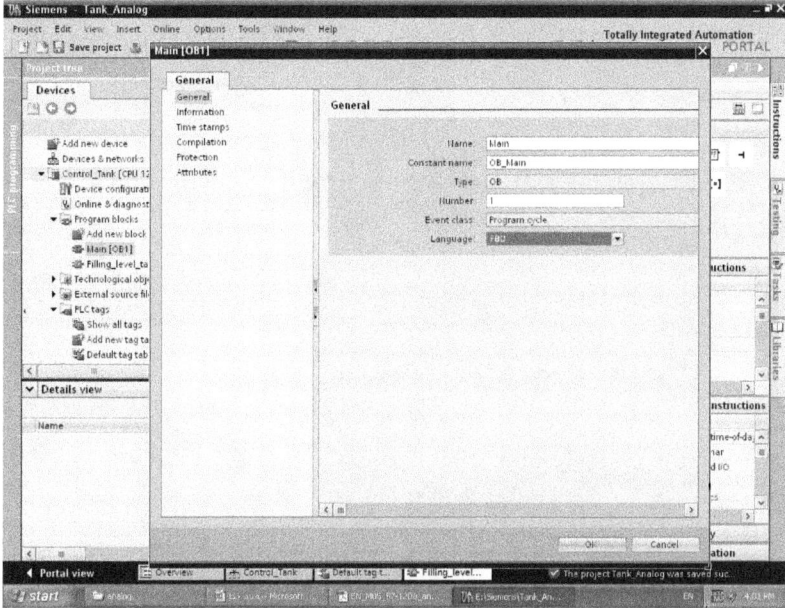

13- Double click Main [OB1] and drag & drop the block Filling_level_tank1 [FC1].

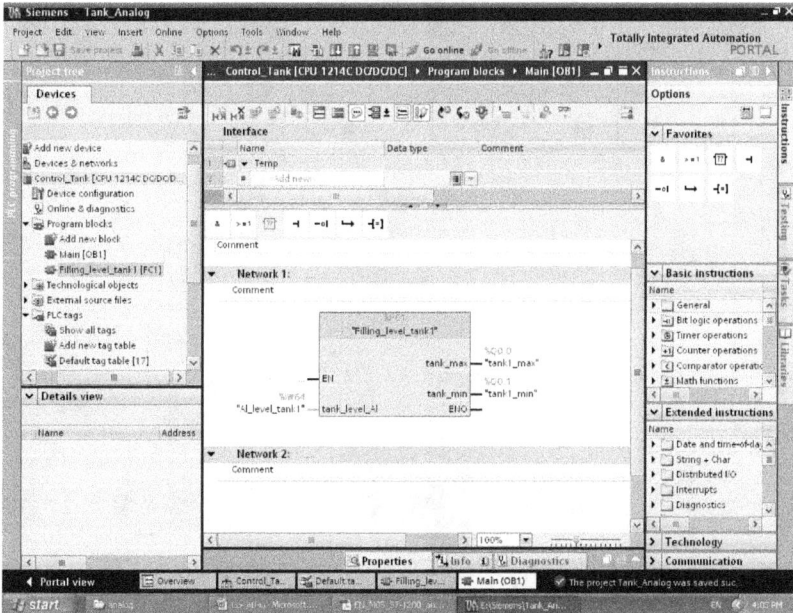

14- Right click CPU 317-2 PN/DP (→ Compile → All)

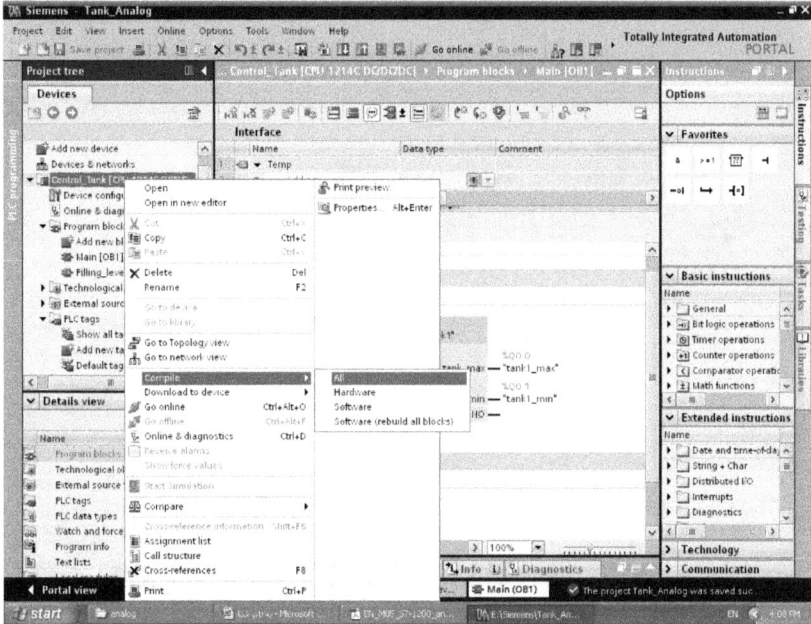

15- Right click CPU 317-2 PN/DP (→ Download to device → All)

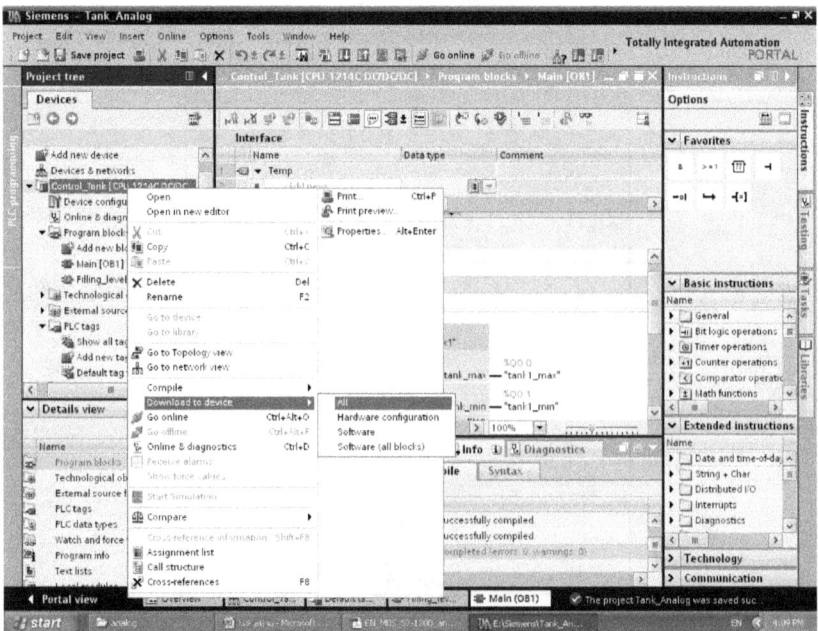

16- Specify the PG/PC interface (→ PG/PC interface for loading → Load)

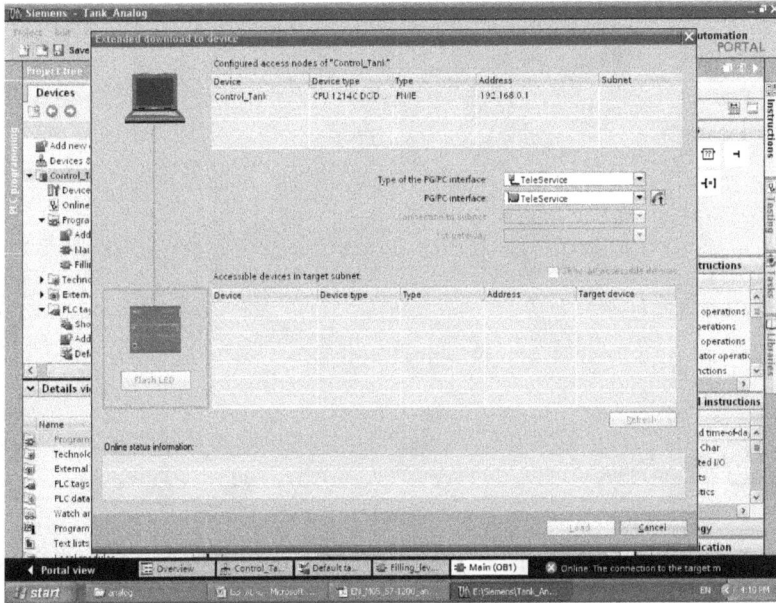

17- Right click CPU 317-2 PN/DP (→ Start Simulation).

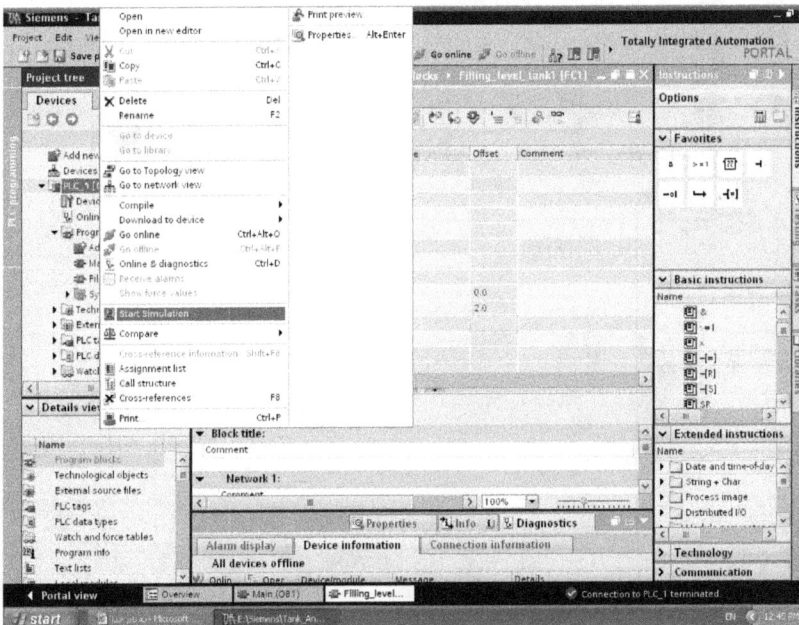

18- Click on 'Load' once more (→ Load).

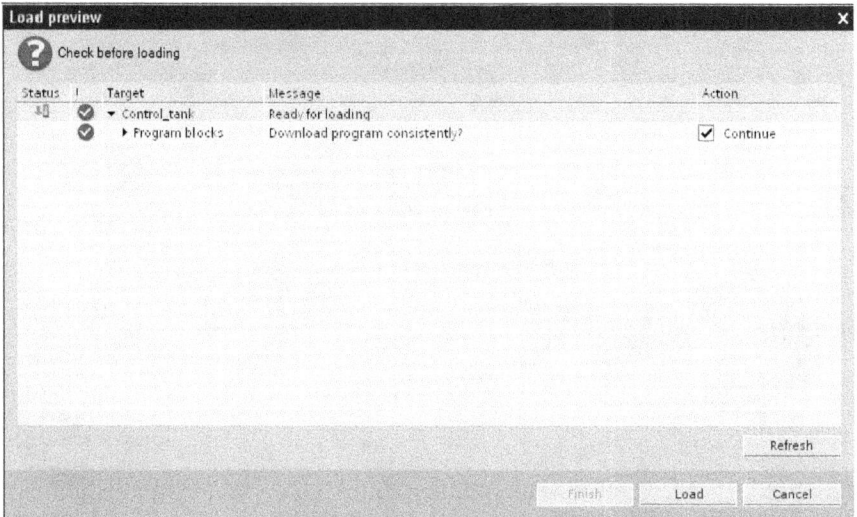

19- Click on the symbol monitoring on/off and write 27600 in IW64.

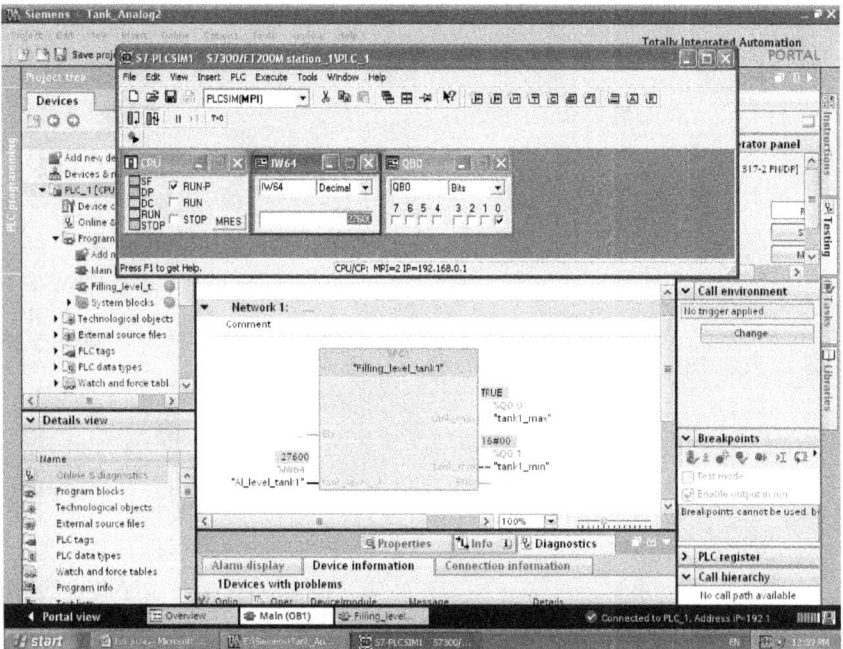

20- Click on the symbol ![symbol] monitoring on/off and write 200 in IW64.

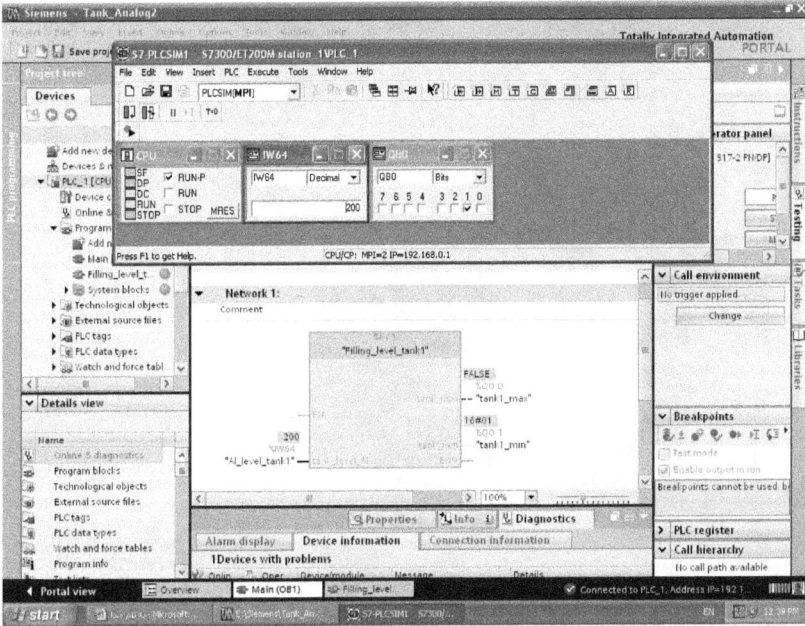

6.3 Working with HMI on the SIMATIC S7 TIA Portal

The following steps explain how to program HMI for the above example.

21- Add new device (→ Project view → Devices & networks → Add new device → HMI → SIMATIC Basic Panel → 6" Display → KTP600 Basic PN → Add)

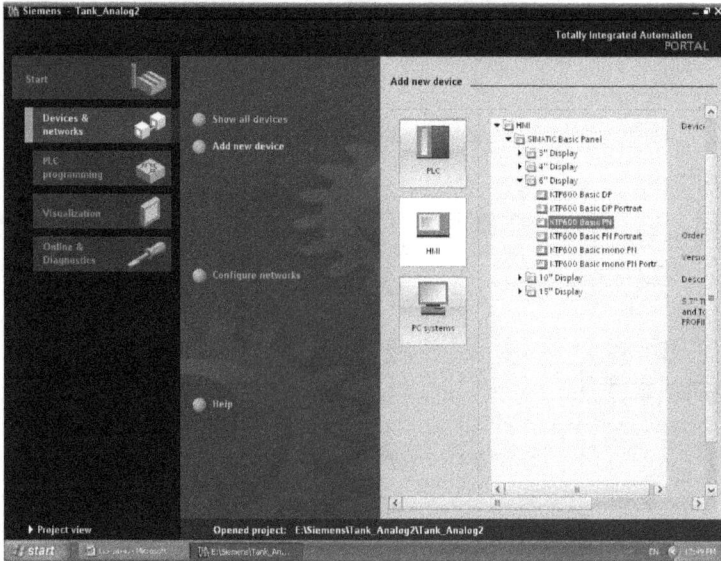

22- HMI Device Wizard (→ PLC connections → Browse → CPU 317-2 PN/DP → Next)

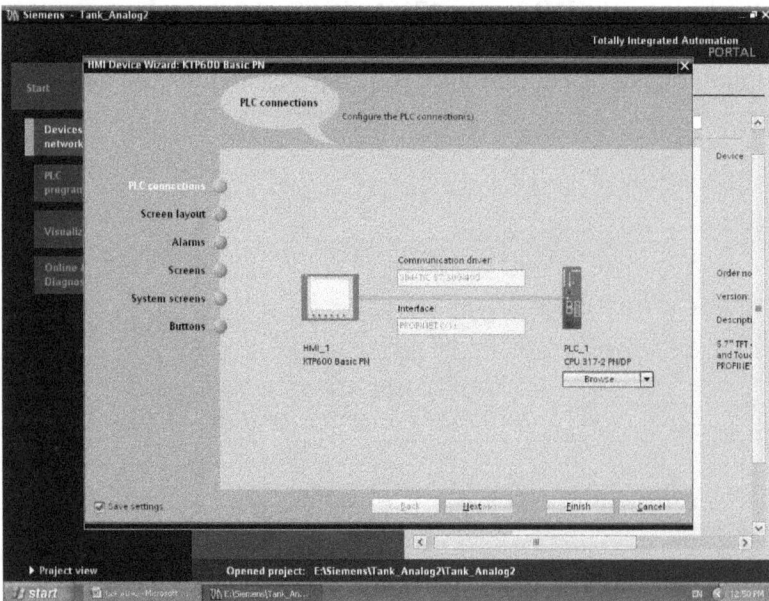

23- HMI Device Wizard (→ Screen layout → Next)

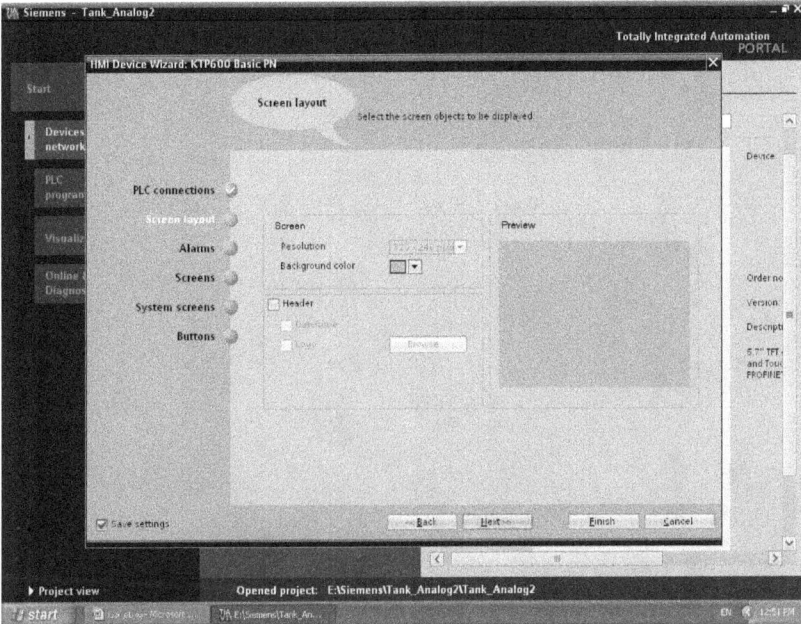

24- HMI Device Wizard (→ Alarms → Next)

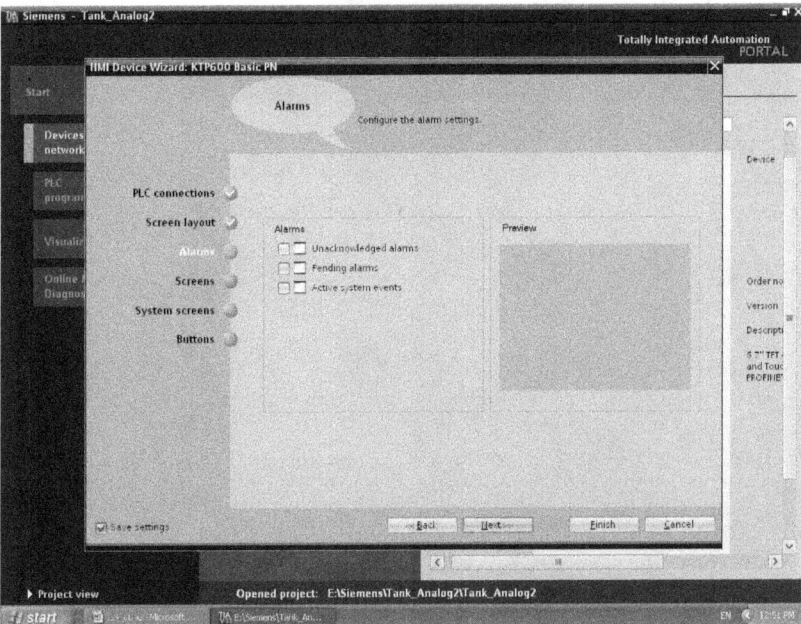

25- HMI Device Wizard (→ Screens → Next)

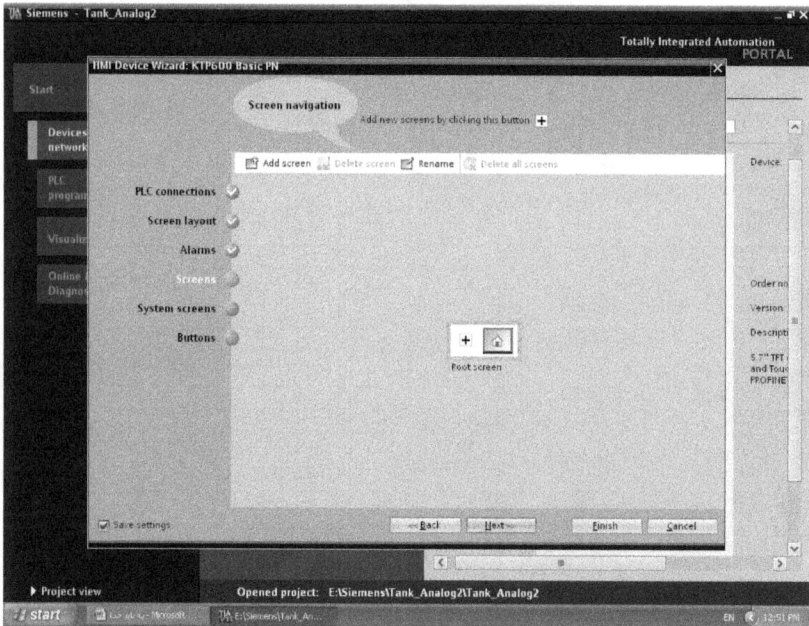

26- HMI Device Wizard (→ System screens → Root screen → System screens → Operating modes → Stop Runtime → Next)

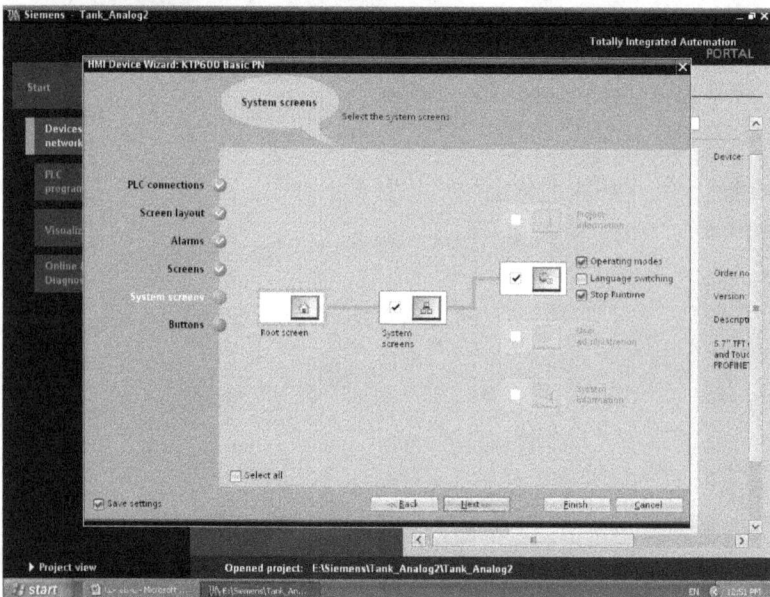

27- HMI Device Wizard (→ Buttons → Finish)

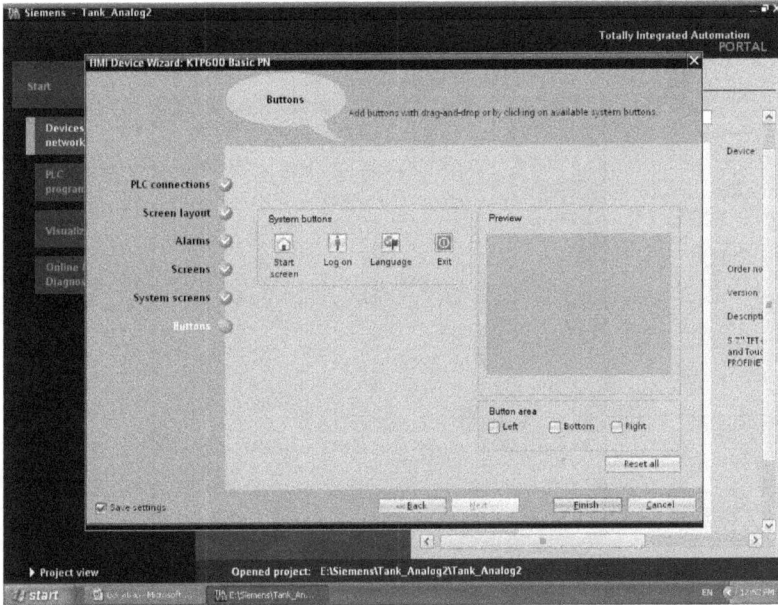

28- Delete the System screens and Welcome to HMI_1 (KTP600 Basic PN)!

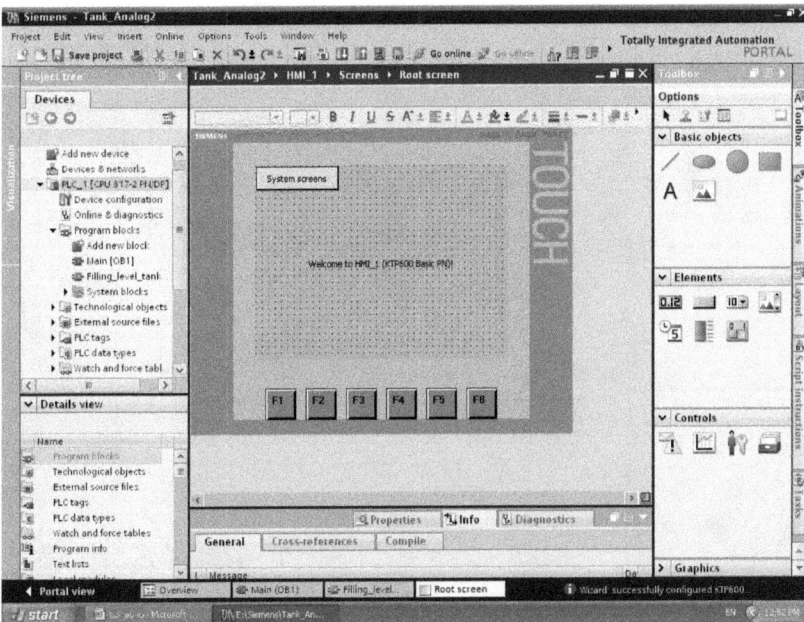

29- Insert a Bar and two Pilot Lights and enter Al_level_tank1 in PLC Tag and 27648 in Static for Bar graph.

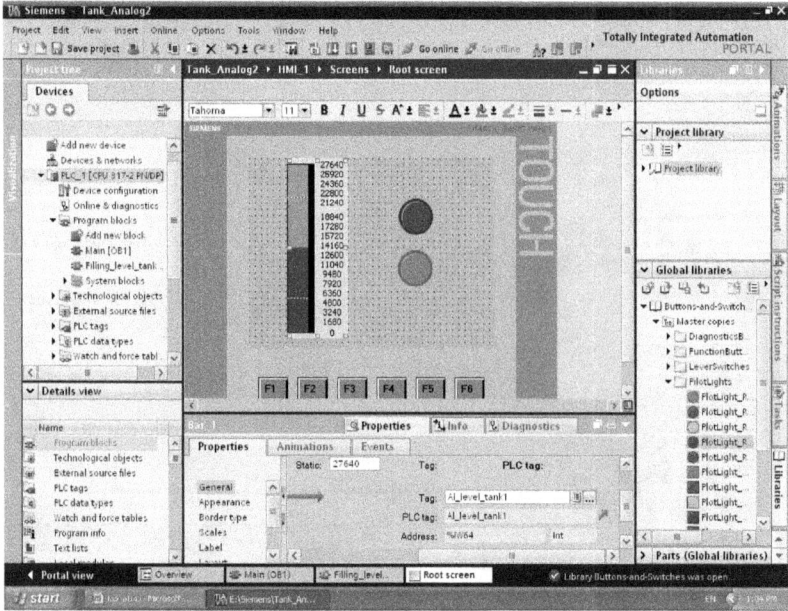

30- Enter tank1_max in Process Tag for PilotLight_Red

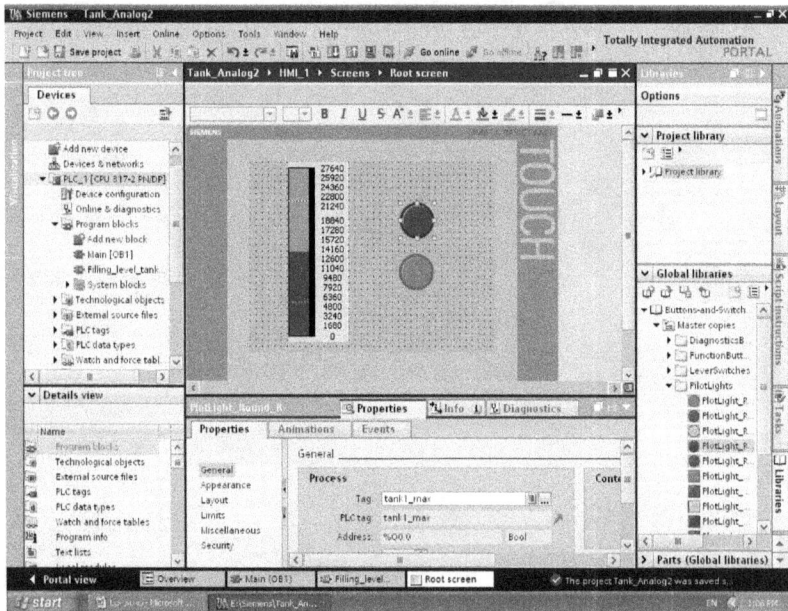

31- Enter tank1_min in Process Tag for PilotLight_Green

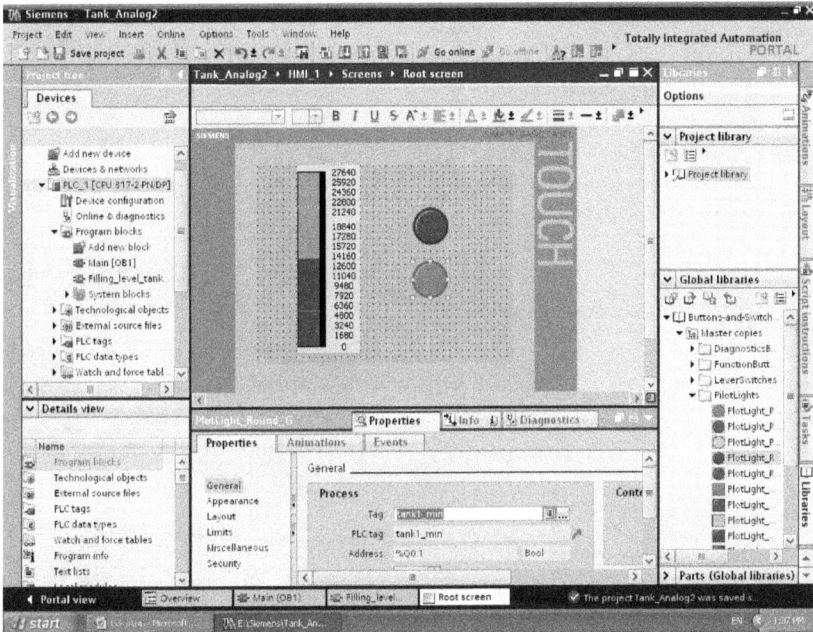

32- Save project → Right click CPU 317-2 PN/DP → Compile → All

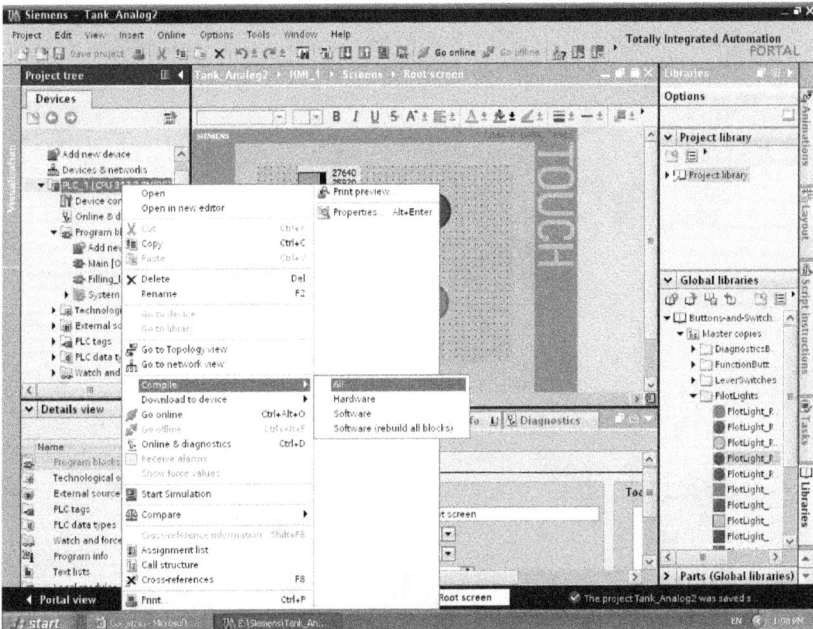

33- Right click CPU 317-2 PN/DP → Start Simulation

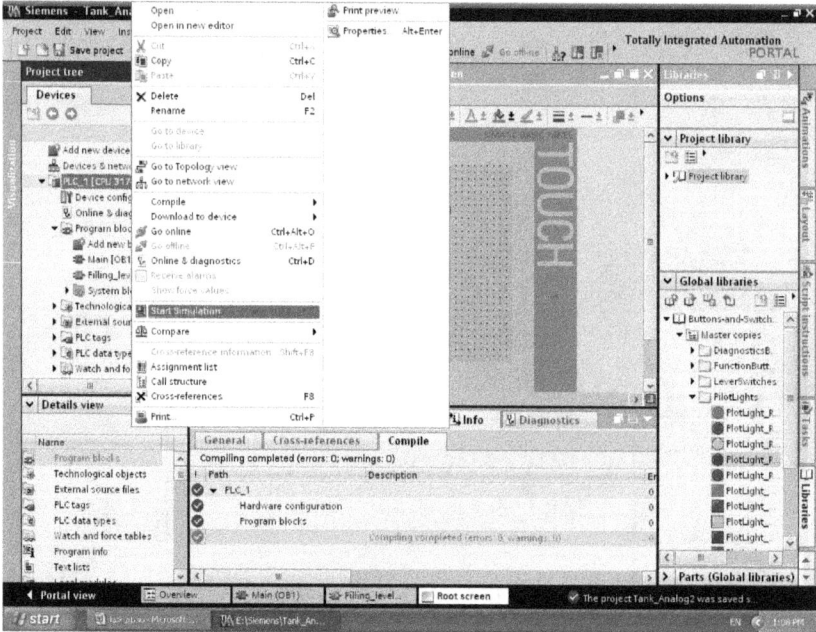

34- Extended download to device → Load

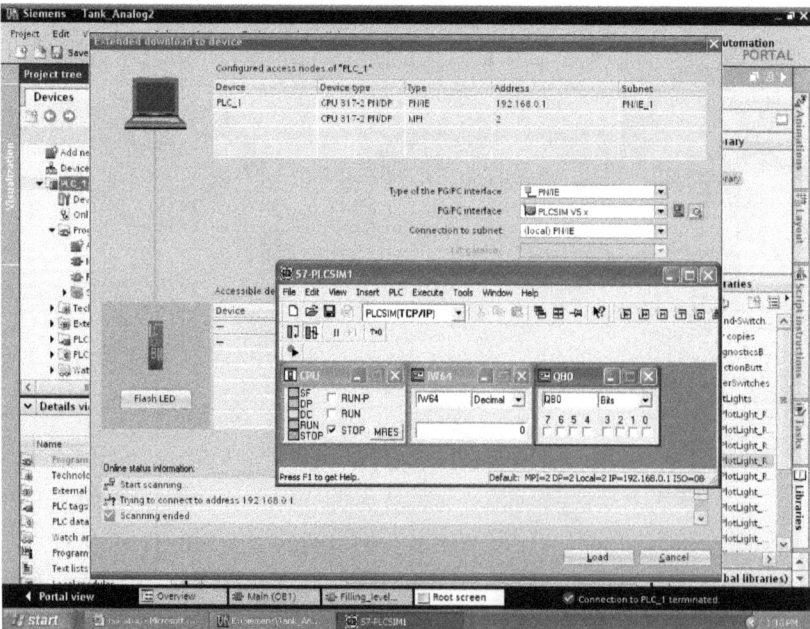

35- Load preview → Load

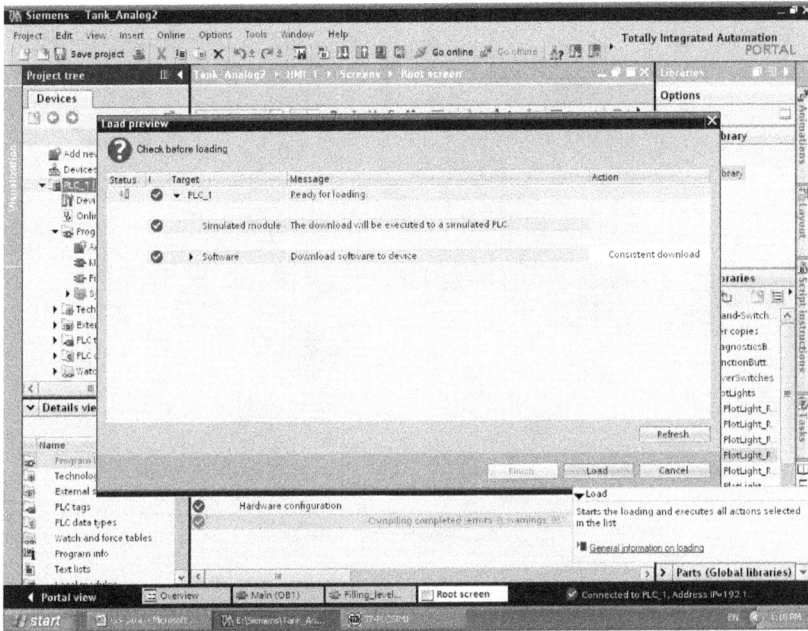

36- Right click HMI_1 (KTP600 Basic PN) → Compile → All

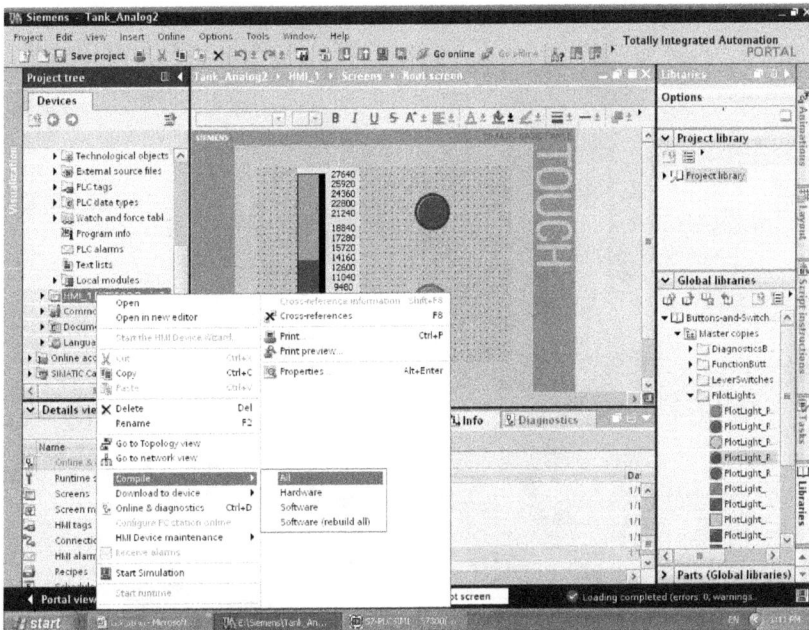

37- Right click HMI_1 (KTP600 Basic PN) → Start Simulation

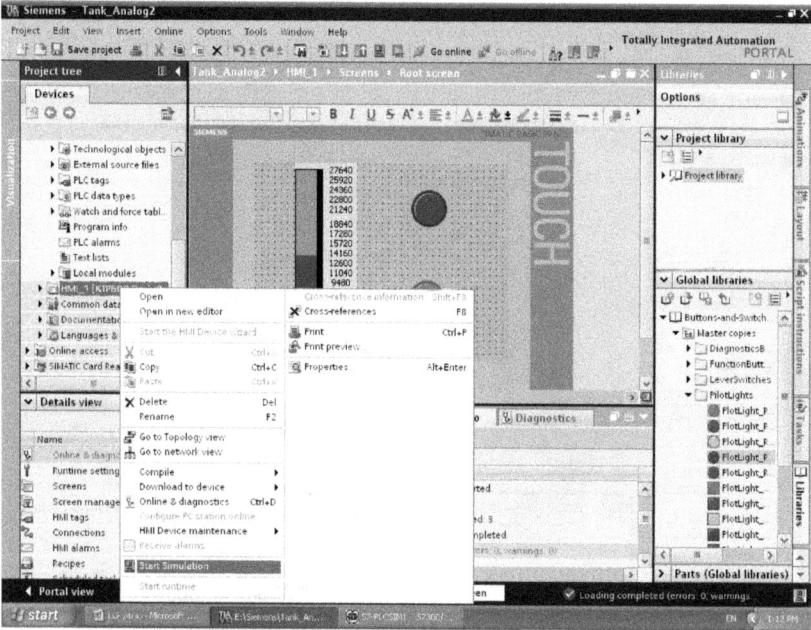

38- Enter 0 in IW64 the result is:

39- Enter 15000 in IW64 the result is:

40- Enter 27400 in IW64 the result is:

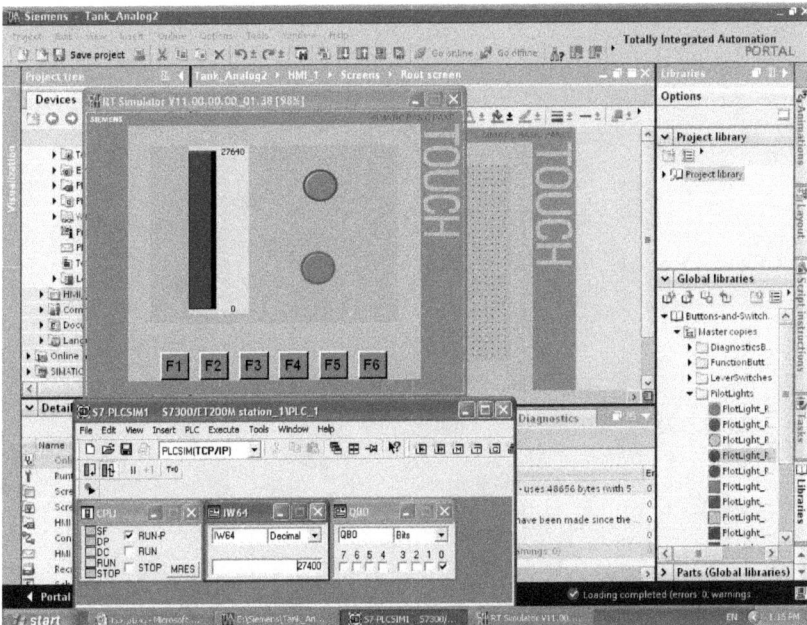

Chapter 7

Implementation of PID and Fuzzy Control in PLC

7.1 Introduction

The main technology for industrial PID control (proportional, integral and derivative) is pneumatic, hydraulic or mechanical, and the controller was easy to adjust manually. PID control is used to control and maintain processes and can be used to control physical variables such as temperature, pressure, and flow rate and tank liquid level. This technique is widely used in industry today to achieve proper control under different process conditions. PID is an equation that the controller uses to evaluate control variables. Process variable (PV) for example temperature is measured and feed backed to the controller. The controller then compares the feedback with the set point (SP) and generates an error value. One or more of the three steps of proportionality, integral, and derivative, meet this value. Therefore, the controller gives the necessary commands and changes the control variable (CV) to correct the error (E). These steps constitute a repetitive process. The following is the application of a common control loop.

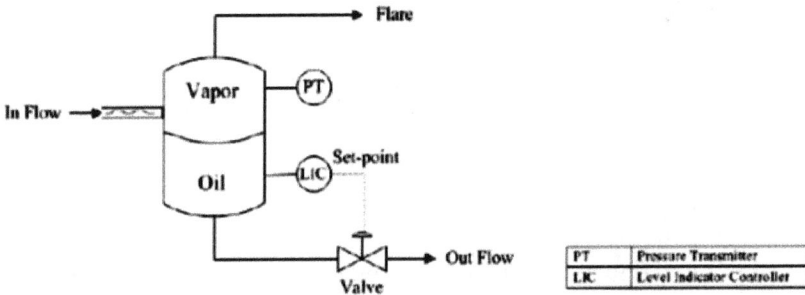

In this example, the oil flows into the tank at an unstable rate. The oil level in the tank is a process variable that is measured and fed to the controller. The operator inserts a set point for a desired level. The controller compares the current level with the set point and produces a value that is evaluated by the PID method. The controller then adjusts the valve position to correct the error.

7.2 PID control theory

PID controllers generally use feedback control loop in industrial and control systems. The controller first calculates the difference between the measured process variable and the desired set point. It then attempts to minimize the error by increasing or decreasing the control inputs or outputs in the process until the process variable moves close to the point. This method is very useful when the mathematical model of process or control is very complex or unknown. To increase performance, for example to increase system responsiveness, PID parameters must be adjusted according to specific application. The block diagram below shows an example of a heating process controlled by a PID controller (PLC). The operator sets the desired temperature as set point. The furnace temperature is measured and fed to the controller. The feedback is compared to the set point and an error value is calculated.

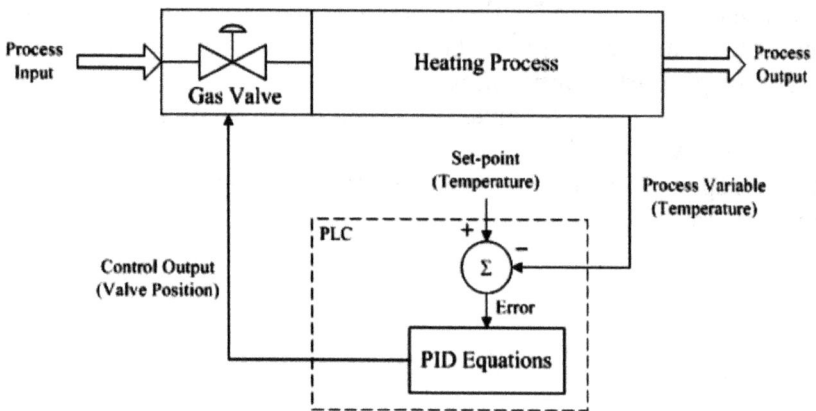

Then, the PID equations determine the appropriate valve position to correct the error. In fact, this is an example of a PID feedback loop. The following figure shows the step response curve after the controller response to a change in set point. One of the advantages of PID is that there is a direct relationship between process responses and the use of the three expressions P, I, and D. There are two steps to designing a PID system. The designer must first select the PID controller structure for example P, PI or PID. Second, the numeric values for the PID parameters must be selected to set the controller. The three parameters for the PID algorithm are proportional, integral, and derivative constants. The proportionality constant responds based on the current error. The reaction is determined by the sum of the current errors in the integral constant and determines the reaction derivative constant using the error rate of change. These three operations are then used to adjust the process by the control element, such as the valve position. Simply put, P is dependent on the current error, I is dependent on the sum of past errors, and D is the prediction of future errors based on the current rate of error change.

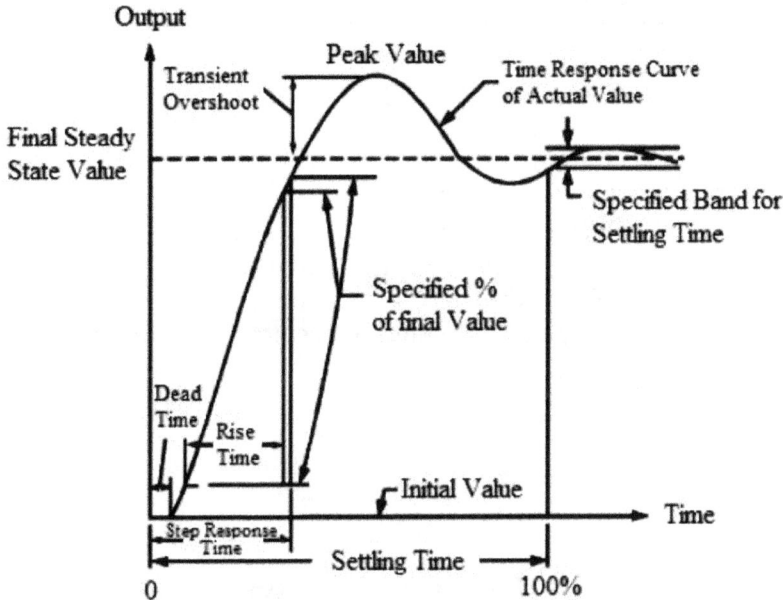

7.2.1 Proportional control

In this controller, we have:

$$P_{out} = K_p \, e$$

Where:

P_{out}: Proportional portion of controller output

K_p: Proportional gain

e: Error signal, e = Set-point − Process Variable

The following figure shows the proportional control. There is always a steady state error in proportional control. The error decreases with increasing gain, but the tendency for oscillation also increases.

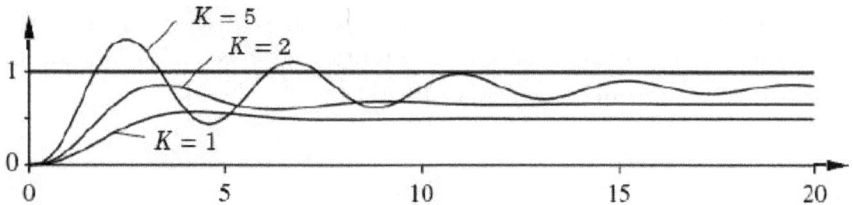

7.2.2 Integral control

Integral control corrects small errors. The integral can be adjusted by adjusting the reset speed, which is a time factor. The shorter the reset speed, the faster the error correction is. However, too slow a reset speed makes the performance unpredictable. In this controller, we have:

$$I_{out} = \frac{1}{T_i} \int e \, dt = K_i \int e \, dt$$

Where:

I_{out}:   Integral portion of controller output

T_i: Integral time, or reset time

K_i: Integral gain

7.2.3 Derivative control

The derivative part of the control output is concerned with the rate of change in the error signal. The derivative induces a larger system response to a faster rate of change than a smaller rate of change. The derivative is set by the coefficient of time called the velocity time. Applying a higher speed time will result in unpredictable overshoot and control. In this controller, we have:

$$D_{out} = T_d \frac{d}{dt} e = K_d \frac{d}{dt} e$$

Where:

D_{out}:   Derivative portion of controller output

T_d: Derivative time

K_d: Derivative gain

e: Error signal, e = Set-point – Process Variable

By applying the three statements above, the control output or control variable is as follows:

$$\text{Control Variable} = P_{out} + I_{out} + D_{out}$$

7.3 Selecting the PID controller structure

Procedure for choosing the structure of the PID controller is shown in the block diagram below.

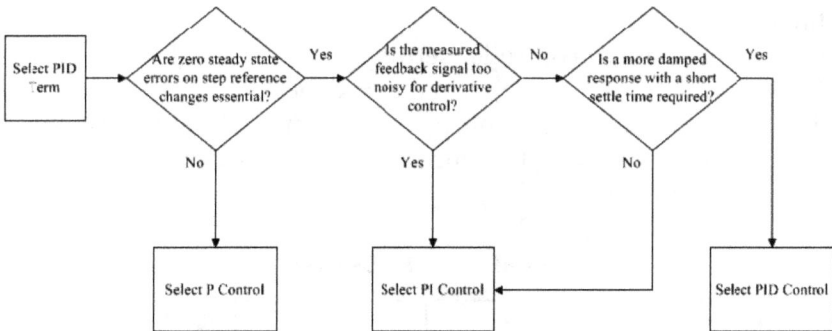

7.4 PID controller settings

Proportional Gain (Kp): Larger proportionality means faster response. However, a larger proportion may cause the process to be oscillated and unstable.

Integral Gain (Ki): Larger integral gains cause steady state errors to be eliminated faster. However, in the larger overshoot mode, the system may be unstable.

Derivative Gain (Kd): Larger derivative Gain reduces overshoot and slows down the transient response and may cause instability due to amplified signal noise in the error derivation.

7.5 Ziegler-Nichols method for tuning the PID controller

The Ziegler-Nichols method is a continuous cycle method for controller tuning. The term continuous cycle is due to a constant oscillation with constant amplitude based on trial and error. The following steps illustrate how to do this.

1. Allow the process to be as constant as possible, turn off the integral mode (set time zero), and then turn off the derivative mode.
2. Assign a small amount to the proportional mode (only controller gain K = 0.5) and set the controller to automatic mode.
3. Make a small set point change so that the control variable is removed from the set point.
4. Increase gin slowly.
5. Repeat steps 3 and 4 until there is a continuous swing called the final gain.
6. Calculate the PID controller settings using the Ziegler-Nichols adjustment relationships shown in the table below.
7. Evaluate the Ziegler-Nichols controller settings by applying a small point set and observing the closed loop response. Make fine adjustments if necessary.

	P Control	**PI Control**	**PID Control**
K_p	0.5 K	0.45 K	0.6 K
K_i		$\dfrac{1.2}{T}$	$\dfrac{2}{T}$
K_d			$\dfrac{T}{8}$

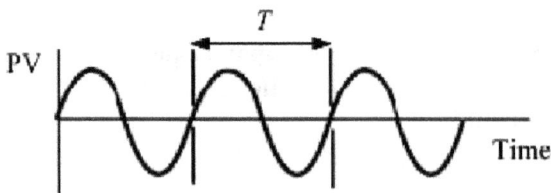

7.6 Cascade control

PID control can be improved by changing PID parameters, improving measurements such as sampling speed, accuracy, accuracy and low pass filter. Another proven method is to use multiple cascade PID controllers. Sequential PID controllers can be used for better dynamic performance. Using the previous example, the sequential control is as follows.

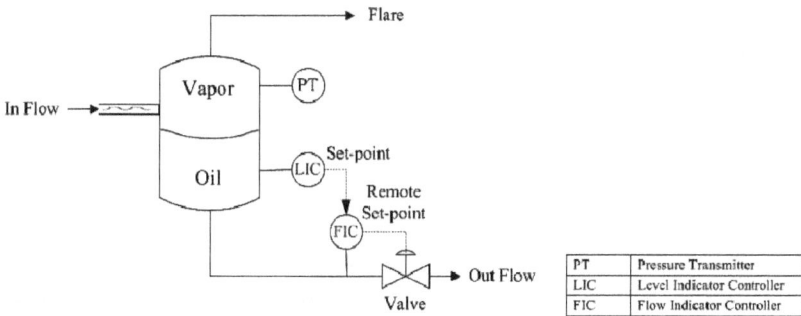

In sequential control there are two PID controllers. One PID controller controls the set point of the other. A PID controller acts as an external loop controller and controls the basic physical parameter such as temperature. The other controller acts as an internal loop controller whose external controller output is called a set point and controls a very fast switching parameter such as the flow rate. The inner ring has a faster response time than the outer ring. The block diagram of the sequential control loop is shown in the following figure.

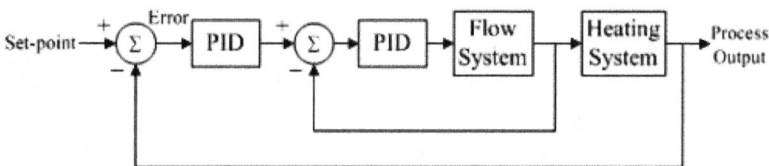

Sequential control is especially useful when the main perturbation is in the inner ring, because in the sequential structure the internal perturbation correction occurs as soon as the secondary sensor detects the perturbation. The three advantages of sequential control are:

1. Disturbances affecting the secondary variable can be corrected by the secondary controller before their impact is felt by the primary variable.

2. Closed-loop control around the secondary part reduces the post-fuzzy process seen by the primary controller, thereby increasing the response speed. Therefore, the sensitivity of the primary process variable to the process perturbations also decreases.

3. The use of a secondary loop reduces the effect of stick control valve or nonlinearity of the operator.

7.7 Fuzzy control

7.7.1 Introduction

A fuzzy logic system is defined as a nonlinear mapping of input data to scalar output data. A fuzzy controller is a nonlinear controller that is described by linguistic rules instead of differential equations. In the fuzzy control system, an expert performs the control strategy. Fuzzy control is more useful than traditional control in the following cases:

1. The controlled process is nonlinear.
2. No mathematical model of the process is available.
3. Expert knowledge plays a key role in process control.
4. A multidimensional nonlinear relationship (control law) should be used to make it easy to understand and improve.

7.7.2 Example (Room Temperature Control)

Room temperature equipped with a hot water heater should be controlled by adjusting the position of the valve in the radiator. A person should apply the following rules:

If the settings are not OK but the changes are in the correct order, keep the current settings with the following more explanation:

If the temperature is too hot but it does not decrease, do not change the valve position. If the temperature is too cold and it decreases, increase the opening of the valve.

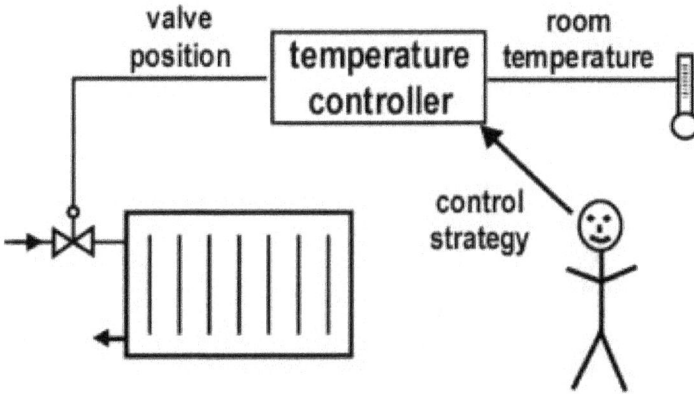

The key concepts for representing knowledge in fuzzy systems are linguistic variables and IF-THEN rules. Interface is the core component of information processing. As a part of a control system, a fuzzy controller processes the numerical inputs to the numerical outputs. Therefore, fuzzification and defuzzification complement the interface.

Fuzzification: Converts the measured numerical values to a fuzzy representation of the input position.

Interface: Converts a fuzzy input representation into a fuzzy decision.

Defuzzification: Converting a fuzzy decision to a real decision.
T
he diagram below shows a fuzzy system. The membership function of fuzzy set A is the actual values between 0 and 1 and is shown below.

$$\mu_A : X \rightarrow [0,1]$$

These fuzzy sets are shown in the figure below.

			TOO WARM	OK	T_{Dev} TOO COLD
y			TOO WARM	OK	TOO COLD
		INCR	R_1: NB	R_2: NS	R_3: ZE
	ΔT	EQUAL	R_4: NS	R_5: ZE	R_6: PS
		DECR	R_7: ZE	R_8: PS	R_9: PB

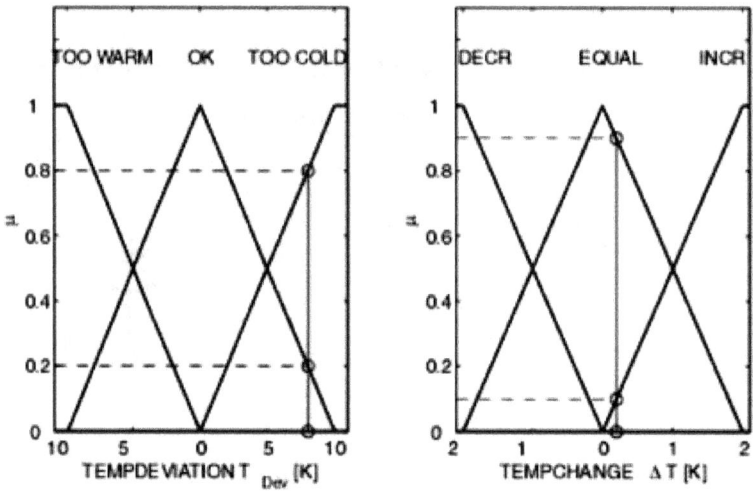

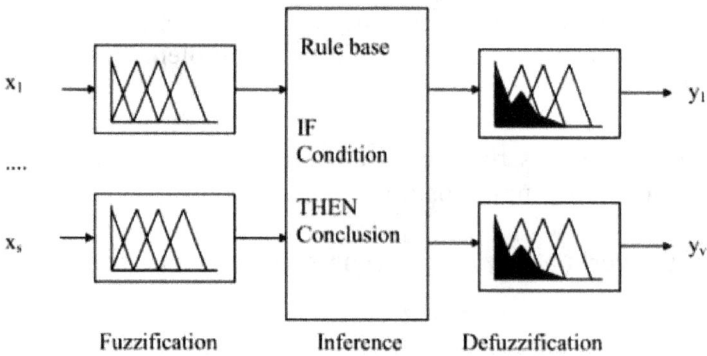

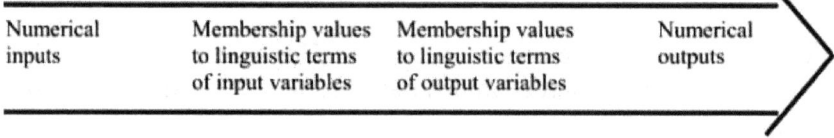

The interface consists of three steps:

1- Aggregation
2- Activation
3- Accumulation

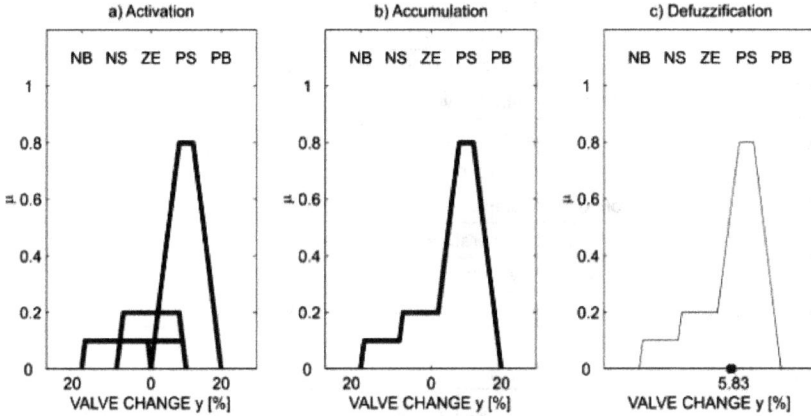

a) Activation   b) Accumulation   c) Defuzzification

7.8 Examples of PID and fuzzy control in PLC

7.8.1 Fuzzy control of oven temperature

Here, we consider the Delta family PLCs. Software for designing Delta family PLCs is WPLSoft. The thermal environment of the stove is the fast heating environment (D13 = K16) and the target temperature is 120°C (D10 = K1200). In order to achieve the best control results of the program, FTC command with GPWM is used for fuzzy temperature control. The DVP04PT-S temperature measurement module is used to measure the existing oven temperature and transfer the result to DVP12SA. After executing the FTC command, the PLC stores the operation results in D22, which is the input to the GPWM command. The GPWM command generates the bandwidth modulation pulses (determined by D22) of output by Y0 to control the heater, thus completing the fuzzy control of the oven temperature.

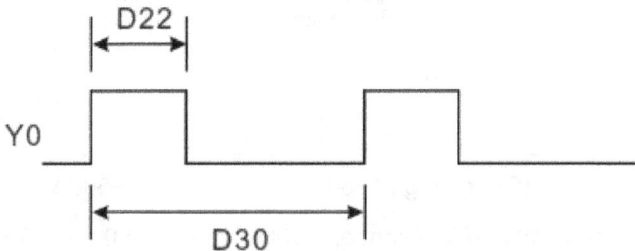

Device	Function
M1	Enabling the execution of FTC instruction
Y0	PWM Pulse output device
D10	Storing the target temperature
D11	Storing the present temperature
D12	Storing FTC sampling time parameter
D13	Storing FTC temperature control parameter
D22	Storing the operation results of FTC instruction
D30	Storing the pulse output cycle of GPWM instruction

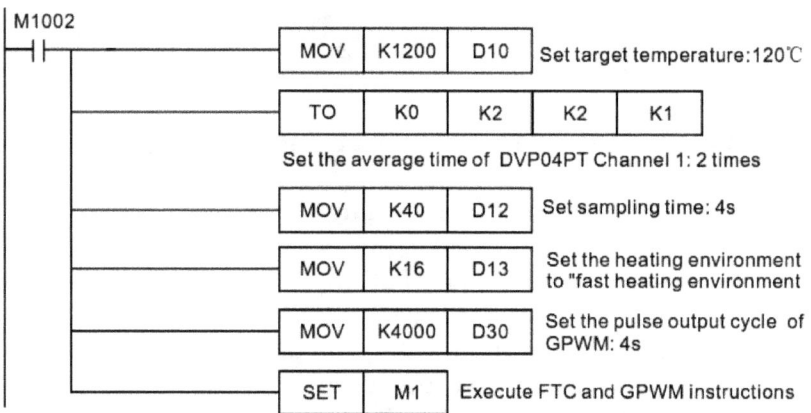

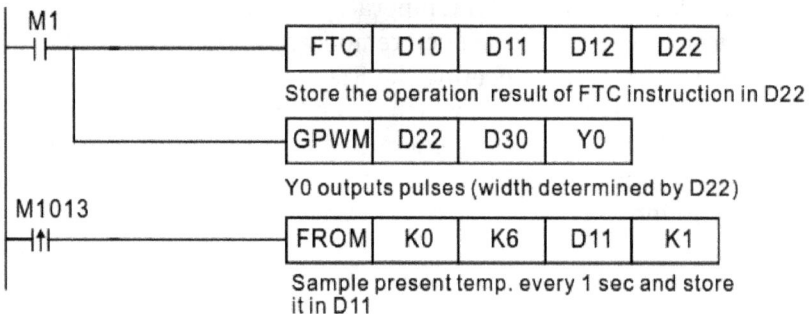

M1002

| MOV | K1200 | D10 | Set target temperature:120℃ |

| TO | K0 | K2 | K2 | K1 |

Set the average time of DVP04PT Channel 1: 2 times

| MOV | K40 | D12 | Set sampling time: 4s |

| MOV | K16 | D13 | Set the heating environment to "fast heating environment |

| MOV | K4000 | D30 | Set the pulse output cycle of GPWM: 4s |

| SET | M1 | Execute FTC and GPWM instructions |

M1

| FTC | D10 | D11 | D12 | D22 |

Store the operation result of FTC instruction in D22

| GPWM | D22 | D30 | Y0 |

Y0 outputs pulses (width determined by D22)

M1013

| FROM | K0 | K6 | D11 | K1 |

Sample present temp. every 1 sec and store it in D11

Format of FTC instruction:

| FTC | S_1 | S_2 | S_3 | D |

$S_1 \rightarrow$ Set value (SV) (Range:1~5000, shown as 0.1°~500°)

$S_2 \rightarrow$ Present value (PV) (Range:1~5000, shown as 0.1°~500°)

$S_3 \rightarrow$ Parameter (Users need to set parameters S_3 and $S_3 + 1$)

$D \rightarrow$ Output value (MV) (Range: 0 ~ S_3*100)

Setting of S_3 and $S_3 + 1$:

Device	Function	Range
S_3	Sampling time (Ts)	1~200 (unit: 100ms)
$S_3 + 1$	b0: temperature unit b1: filter function b2: heating environment b3~b15: reserved	b0 = 0 means ℃ ; b0 = 1 means ℉
		b1 = 0 means without filter function; b1 = 1 means with filter function
		b2 = 1 Slow heating environment
		b3 = 1 General heating environment
		b4 = 1 Fast heating environment
		b5 = 1 High-speed heating environment

7.8.2 Level control by FGS (fuzzy gain scheduling)

The control of the FGS level is shown below.

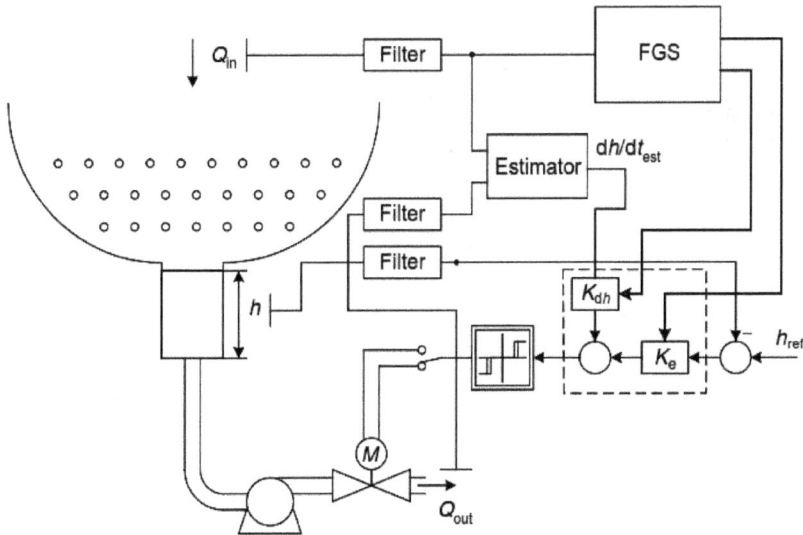

The FGS has two Q_{in} vapor and a ΔQ_{in} vapor flow as inputs and two K_{dh} and K_e as outputs. The FGS determines the new values of the Gain coefficients using the fuzzy law table with nine fuzzy laws. The inputs are three fuzzy sets with triangular membership functions while singleton columns represent the outputs. The output is calculated according to the COG (center of gravity) method as follows:

$$u_{FC}(x_k, y_k) = \frac{\sum_{i=1}^{r} u_{c_i} \cdot \mu_{FR^i}(x_k, y_k, u_{c_i})}{\sum_{i=1}^{r} \mu_{FR^i}(x_k, y_k, u_{c_i})}$$

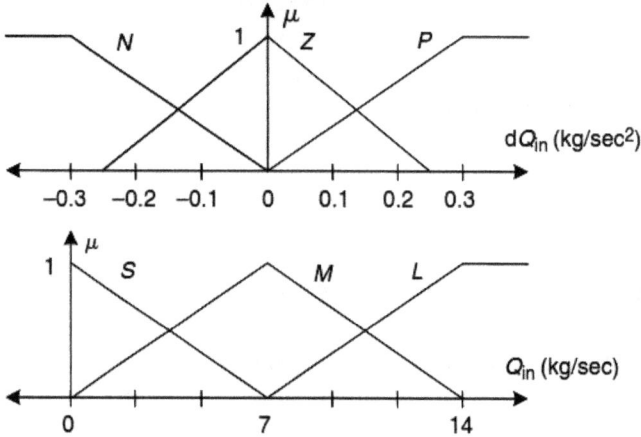

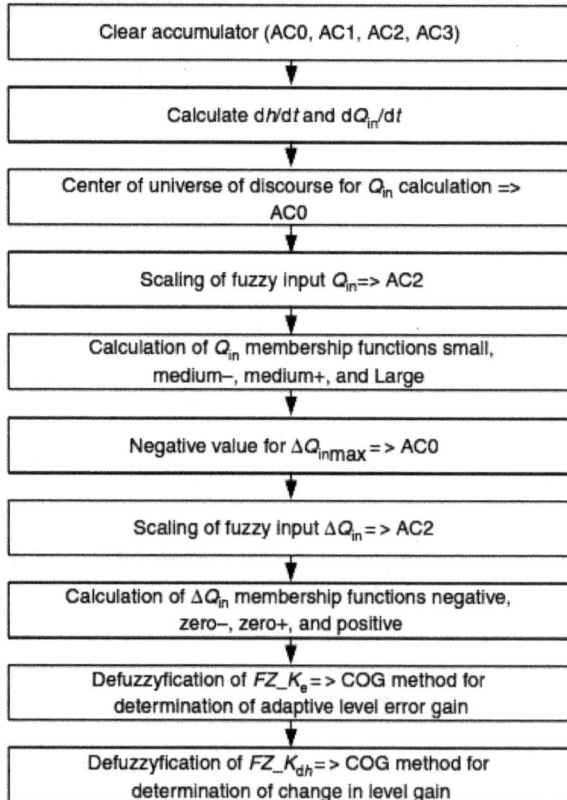

FGS algorithm

Clear accumulator (AC0, AC1, AC2, AC3)

↓

Calculate dh/dt and dQ_{in}/dt

↓

Center of universe of discourse for Q_{in} calculation => AC0

↓

Scaling of fuzzy input Q_{in} => AC2

↓

Calculation of Q_{in} membership functions small, medium−, medium+, and Large

↓

Negative value for $\Delta Q_{in_{max}}$ => AC0

↓

Scaling of fuzzy input ΔQ_{in} => AC2

↓

Calculation of ΔQ_{in} membership functions negative, zero−, zero+, and positive

↓

Defuzzyfication of FZ_K_e => COG method for determination of adaptive level error gain

↓

Defuzzyfication of FZ_K_{dh} => COG method for determination of change in level gain

Estimator calculates the dh/dt level based on flow measurement. The calculated value is multiplied by the K_{dh} and summed with the error signal between the reference level h_{ref} and the level measured h. All measured variables are filtered by a first-order filter.

The structure of the PLC program that implements the FGS control algorithm is illustrated in the following figure, which is implemented with Step 7.

Network 1   clear accumulator

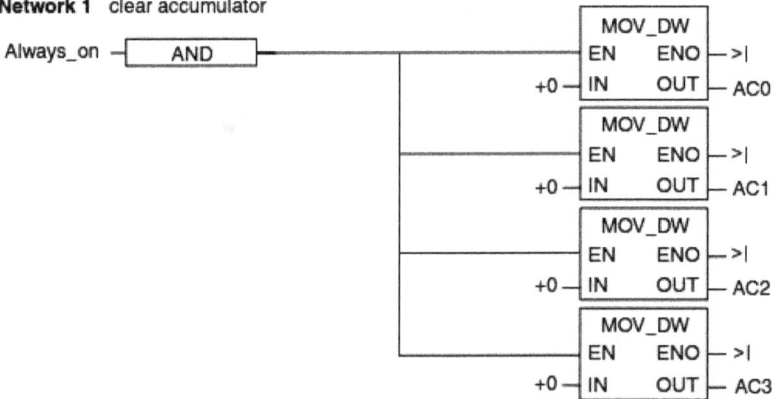

```
                                          ┌── MOV_DW ──┐
Always_on ──┤    AND    ├─────────┬──────┤ EN    ENO ├─>|
                                  │  +0 ─┤ IN    OUT ├─ AC0
                                  │       └────────────┘
                                  │       ┌── MOV_DW ──┐
                                  ├──────┤ EN    ENO ├─>|
                                  │  +0 ─┤ IN    OUT ├─ AC1
                                  │       └────────────┘
                                  │       ┌── MOV_DW ──┐
                                  ├──────┤ EN    ENO ├─>|
                                  │  +0 ─┤ IN    OUT ├─ AC2
                                  │       └────────────┘
                                  │       ┌── MOV_DW ──┐
                                  └──────┤ EN    ENO ├─>|
                                     +0 ─┤ IN    OUT ├─ AC3
                                          └────────────┘
```

Network 2   dh/dt and dQ_in/dt calculation

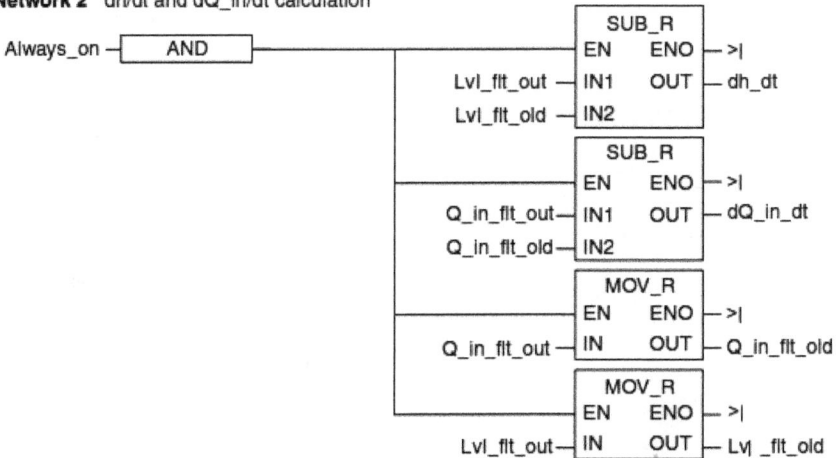

```
                                               ┌─── SUB_R ───┐
Always_on ──┤    AND    ├─────────┬───────────┤ EN     ENO ├─>|
                                  │ Lvl_flt_out ─┤ IN1    OUT ├─ dh_dt
                                  │ Lvl_flt_old ─┤ IN2         │
                                  │             └─────────────┘
                                  │             ┌─── SUB_R ───┐
                                  ├────────────┤ EN     ENO ├─>|
                                  │Q_in_flt_out─┤ IN1    OUT ├─ dQ_in_dt
                                  │Q_in_flt_old─┤ IN2         │
                                  │             └─────────────┘
                                  │             ┌─── MOV_R ───┐
                                  ├────────────┤ EN     ENO ├─>|
                                  │Q_in_flt_out─┤ IN     OUT ├─ Q_in_flt_old
                                  │             └─────────────┘
                                  │             ┌─── MOV_R ───┐
                                  └────────────┤ EN     ENO ├─>|
                                   Lvl_flt_out─┤ IN     OUT ├─ Lvl _flt_old
                                                └─────────────┘
```

Network 3   Center of universe of discourse for Q_in calculation => AC0

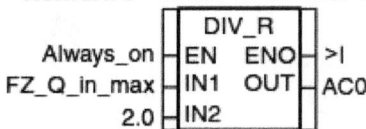

```
                 ┌──── DIV_R ────┐
  Always_on ─────┤ EN      ENO ├─>|
 FZ_Q_in_max ────┤ IN1     OUT ├─ AC0
         2.0 ────┤ IN2          │
                 └───────────────┘
```

Network 4 fuzzy input scaling => AC2

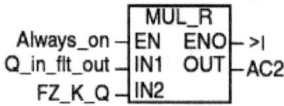

```
                    MUL_R
Always_on ──┤EN    ENO├─ >|
Q_in_flt_out ┤IN1   OUT├─ AC2
   FZ_K_Q ──┤IN2
```

Network 5   Q_in memb func small calculation

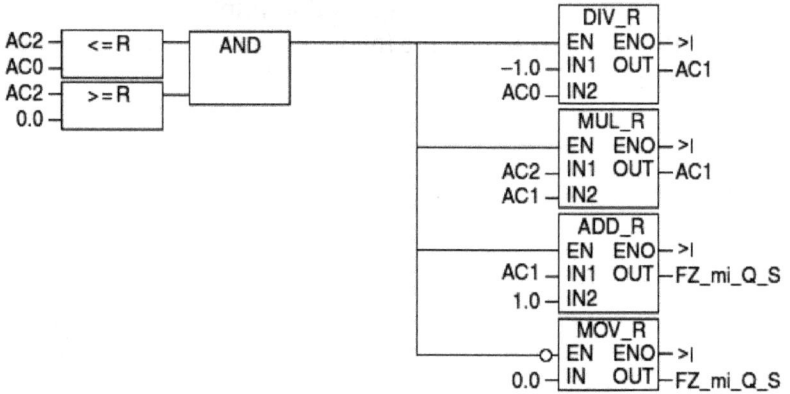

```
AC2 ─┐ ┌──────┐                              DIV_R
ACO ─┤ │ <=R  │  ┌──────┐                  ┌EN    ENO├─ >|
     │ └──────┘  │ AND  ├──────────┐  -1.0─┤IN1   OUT├─ AC1
AC2 ─┐ ┌──────┐  │      │          │   ACO─┤IN2
0.0 ─┤ │ >=R  │  └──────┘          │        MUL_R
     │ └──────┘                    │      ┌EN    ENO├─ >|
                                   ├  AC2─┤IN1   OUT├─ AC1
                                   │  AC1─┤IN2
                                   │        ADD_R
                                   │      ┌EN    ENO├─ >|
                                   ├  AC1─┤IN1   OUT├─ FZ_mi_Q_S
                                   │  1.0─┤IN2
                                   │        MOV_R
                                   └─────o┤EN    ENO├─ >|
                                      0.0─┤IN    OUT├─ FZ_mi_Q_S
```

Network 6   Q_in memb func medium– calculation

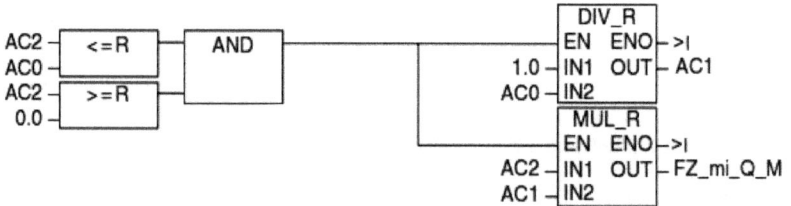

```
AC2 ─┐ ┌──────┐                              DIV_R
ACO ─┤ │ <=R  │  ┌──────┐                  ┌EN    ENO├─ >|
     │ └──────┘  │ AND  ├──────────┐   1.0─┤IN1   OUT├─ AC1
AC2 ─┐ ┌──────┐  │      │          │   ACO─┤IN2
0.0 ─┤ │ >=R  │  └──────┘          │        MUL_R
     │ └──────┘                    │      ┌EN    ENO├─ >|
                                   └  AC2─┤IN1   OUT├─ FZ_mi_Q_M
                                      AC1─┤IN2
```

Network 7   Q_in memb func Medium+ calculation

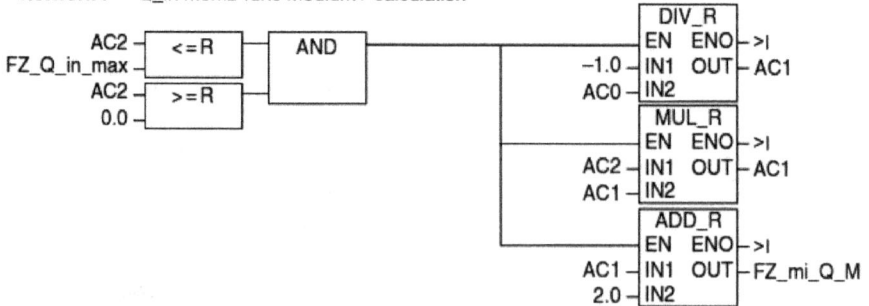

```
       AC2 ─┐ ┌──────┐                          DIV_R
FZ_Q_in_max ┤ │ <=R  │  ┌──────┐              ┌EN    ENO├─ >|
            │ └──────┘  │ AND  ├──────────┐-1.0┤IN1   OUT├─ AC1
       AC2 ─┐ ┌──────┐  │      │          │ ACO┤IN2
       0.0 ─┤ │ >=R  │  └──────┘          │      MUL_R
            │ └──────┘                    │    ┌EN    ENO├─ >|
                                          ├ AC2┤IN1   OUT├─ AC1
                                          │ AC1┤IN2
                                          │      ADD_R
                                          │    ┌EN    ENO├─ >|
                                          └ AC1┤IN1   OUT├─ FZ_mi_Q_M
                                            2.0┤IN2
```

Network 8 Q_in memb func Large calculation

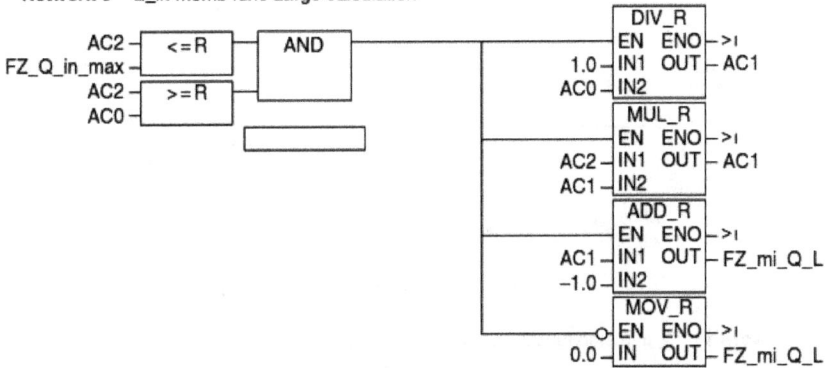

```
AC2 ──┬──< =R──┐        ┌─────────┐                    ┌──────────┐
      │        ├─ AND ──┤         │                    │  DIV_R   │
FZ_Q_in_max ──┘         └─────────┘                    │ EN   ENO ├─ >ı
AC2 ──┬──> =R──┐                                   1.0 ─┤ IN1  OUT ├─ AC1
      │        │                                   AC0 ─┤ IN2      │
ACO ──┘        │                                        └──────────┘
               └────────┐                               ┌──────────┐
                        │                               │  MUL_R   │
                        │                               │ EN   ENO ├─ >ı
                                                   AC2 ─┤ IN1  OUT ├─ AC1
                                                   AC1 ─┤ IN2      │
                                                        └──────────┘
                                                        ┌──────────┐
                                                        │  ADD_R   │
                                                        │ EN   ENO ├─ >ı
                                                   AC1 ─┤ IN1  OUT ├─ FZ_mi_Q_L
                                                  -1.0 ─┤ IN2      │
                                                        └──────────┘
                                                        ┌──────────┐
                                                        │  MOV_R   │
                                                      ○─┤ EN   ENO ├─ >ı
                                                   0.0 ─┤ IN   OUT ├─ FZ_mi_Q_L
                                                        └──────────┘
```

Network 9 Q_in memb func Large calculation

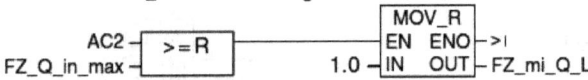

```
AC2 ──┬──> =R──┐              ┌──────────┐
      │        │              │  MOV_R   │
FZ_Q_in_max ──┘              │ EN   ENO ├─ >ı
                        1.0 ─┤ IN   OUT ├─ FZ_mi_Q_L
                              └──────────┘
```

Network 10 Negative value of dQ_in_max calculation = > ACO

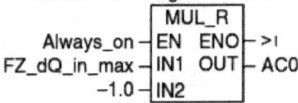

```
                 ┌──────────┐
                 │  MUL_R   │
Always_on ──────┤ EN   ENO ├─ >ı
FZ_dQ_in_max ──┤ IN1  OUT ├─ ACO
       -1.0 ────┤ IN2      │
                 └──────────┘
```

Network 11 Fuzzy input scaling => AC2

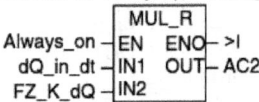

```
               ┌──────────┐
               │  MUL_R   │
Always_on ────┤ EN   ENO ├─ >ı
 dQ_in_dt ────┤ IN1  OUT ├─ AC2
 FZ_K_dQ ─────┤ IN2      │
               └──────────┘
```

Network 12 dQ_in memb func Negative calculation

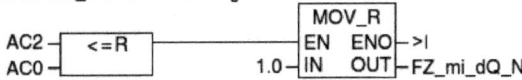

```
AC2 ──┬──< =R──┐              ┌──────────┐
      │        │              │  MOV_R   │
ACO ──┘                       │ EN   ENO ├─ >ı
                        1.0 ─┤ IN   OUT ├─ FZ_mi_dQ_N
                              └──────────┘
```

Network 13 dQ_in memb func Negative calculation

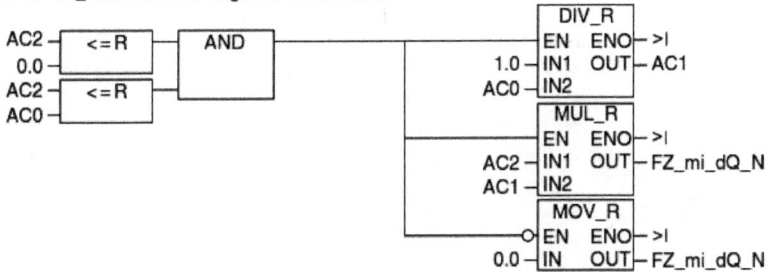

```
AC2 ──┬──< =R──┐        ┌─────────┐                    ┌──────────┐
      │        ├─ AND ──┤         │                    │  DIV_R   │
0.0 ──┘        └─────────┘                    │ EN   ENO ├─ >ı
AC2 ──┬──< =R──┐                                   1.0 ─┤ IN1  OUT ├─ AC1
      │        │                                   ACO ─┤ IN2      │
ACO ──┘        │                                        └──────────┘
               └────────┐                               ┌──────────┐
                        │                               │  MUL_R   │
                        │                               │ EN   ENO ├─ >ı
                                                   AC2 ─┤ IN1  OUT ├─ FZ_mi_dQ_N
                                                   AC1 ─┤ IN2      │
                                                        └──────────┘
                                                        ┌──────────┐
                                                        │  MOV_R   │
                                                      ○─┤ EN   ENO ├─ >ı
                                                   0.0 ─┤ IN   OUT ├─ FZ_mi_dQ_N
                                                        └──────────┘
```

Network 14 dQ_in memb func Zero– calculation

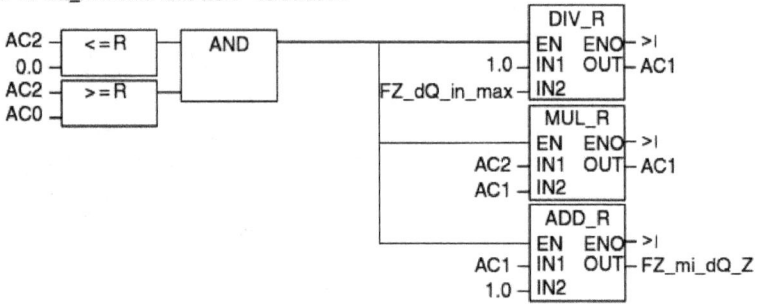

```
AC2 ─┐                                                    ┌──DIV_R──┐
     │ <=R ├──┐ ┌──────┐                                  │EN   ENO ├─>I
0.0 ─┘      ├─┤ AND  ├──────────────────────────┐    1.0─┤IN1  OUT ├─AC1
AC2 ─┐      │ └──────┘                           │ FZ_dQ_in_max─┤IN2      │
     │ >=R ├──┘                                  │    └─────────┘
AC0 ─┘                                           │    ┌──MUL_R──┐
                                                 │    │EN   ENO ├─>I
                                                 ├────AC2─┤IN1  OUT ├─AC1
                                                 │    AC1─┤IN2      │
                                                 │    └─────────┘
                                                 │    ┌──ADD_R──┐
                                                 │    │EN   ENO ├─>I
                                                 └────AC1─┤IN1  OUT ├─FZ_mi_dQ_Z
                                                      1.0─┤IN2      │
                                                      └─────────┘
```

Network 15 dQ_in memb func Zero+ calculation

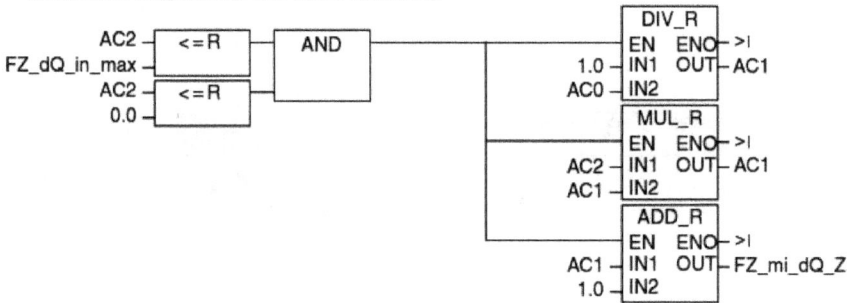

```
       AC2 ─┐                                              ┌──DIV_R──┐
            │ <=R ├──┐ ┌──────┐                            │EN   ENO ├─>I
FZ_dQ_in_max─┘      ├─┤ AND  ├──────────────────────┐  1.0─┤IN1  OUT ├─AC1
       AC2 ─┐       │ └──────┘                       │  AC0─┤IN2      │
            │ <=R ├──┘                               │    └─────────┘
       0.0 ─┘                                        │    ┌──MUL_R──┐
                                                     │    │EN   ENO ├─>I
                                                     ├────AC2─┤IN1  OUT ├─AC1
                                                     │    AC1─┤IN2      │
                                                     │    └─────────┘
                                                     │    ┌──ADD_R──┐
                                                     │    │EN   ENO ├─>I
                                                     └────AC1─┤IN1  OUT ├─FZ_mi_dQ_Z
                                                          1.0─┤IN2      │
                                                          └─────────┘
```

Network 16   dQ_in memb func Positive calculation

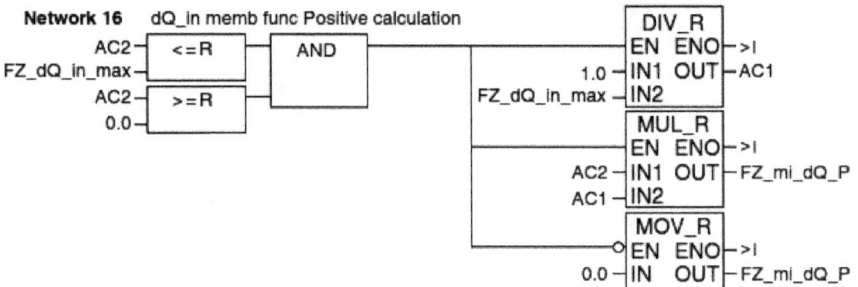

```
       AC2 ─┐                                              ┌──DIV_R──┐
            │ <=R ├──┐ ┌──────┐                            │EN   ENO ├─>I
FZ_dQ_in_max─┘      ├─┤ AND  ├──────────────────────┐  1.0─┤IN1  OUT ├─AC1
       AC2 ─┐       │ └──────┘                       │ FZ_dQ_in_max─┤IN2│
            │ >=R ├──┘                               │    └─────────┘
       0.0 ─┘                                        │    ┌──MUL_R──┐
                                                     │    │EN   ENO ├─>I
                                                     ├────AC2─┤IN1  OUT ├─FZ_mi_dQ_P
                                                     │    AC1─┤IN2      │
                                                     │    └─────────┘
                                                     │    ┌──MOV_R──┐
                                                     └───○│EN   ENO ├─>I
                                                      0.0─┤IN   OUT ├─FZ_mi_dQ_P
                                                          └─────────┘
```

Network 17   dQ_in memb func Positive calculation

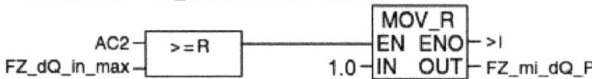

```
       AC2 ─┐                ┌──MOV_R──┐
            │ >=R ├──────────│EN   ENO ├─>I
FZ_dQ_in_max─┘           1.0─┤IN   OUT ├─FZ_mi_dQ_P
                             └─────────┘
```

Network 18   Defuzzyfication FZ_Ke

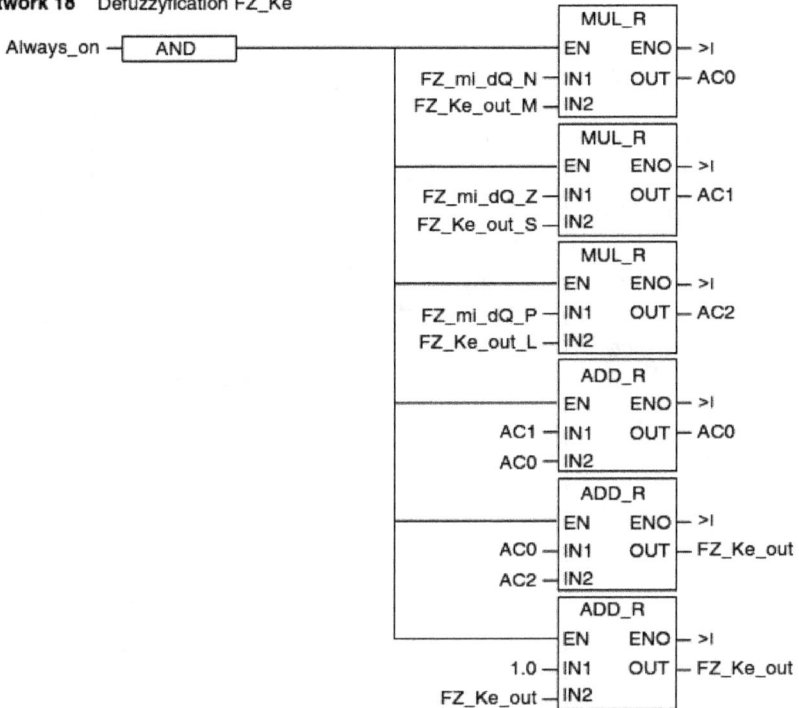

```
Always_on ─┤ AND ├────────────────────┐
                                       │   ┌─────────────────┐
                                       │   │     MUL_R        │
                                       ├───┤EN        ENO├─ >I
                                       │   │              │
                          FZ_mi_dQ_N ──┤IN1       OUT├─ AC0
                          FZ_Ke_out_M ─┤IN2           │
                                       │   └─────────────────┘
                                       │   ┌─────────────────┐
                                       │   │     MUL_R        │
                                       ├───┤EN        ENO├─ >I
                          FZ_mi_dQ_Z ──┤IN1       OUT├─ AC1
                          FZ_Ke_out_S ─┤IN2           │
                                       │   └─────────────────┘
                                       │   ┌─────────────────┐
                                       │   │     MUL_R        │
                                       ├───┤EN        ENO├─ >I
                          FZ_mi_dQ_P ──┤IN1       OUT├─ AC2
                          FZ_Ke_out_L ─┤IN2           │
                                       │   └─────────────────┘
                                       │   ┌─────────────────┐
                                       │   │     ADD_R        │
                                       ├───┤EN        ENO├─ >I
                                 AC1 ──┤IN1       OUT├─ AC0
                                 AC0 ──┤IN2           │
                                       │   └─────────────────┘
                                       │   ┌─────────────────┐
                                       │   │     ADD_R        │
                                       ├───┤EN        ENO├─ >I
                                 AC0 ──┤IN1       OUT├─ FZ_Ke_out
                                 AC2 ──┤IN2           │
                                       │   └─────────────────┘
                                       │   ┌─────────────────┐
                                       │   │     ADD_R        │
                                       └───┤EN        ENO├─ >I
                                 1.0 ──┤IN1       OUT├─ FZ_Ke_out
                          FZ_Ke_out ──┤IN2           │
                                           └─────────────────┘
```

Network 19   defuzzyfication FZ_Kdh

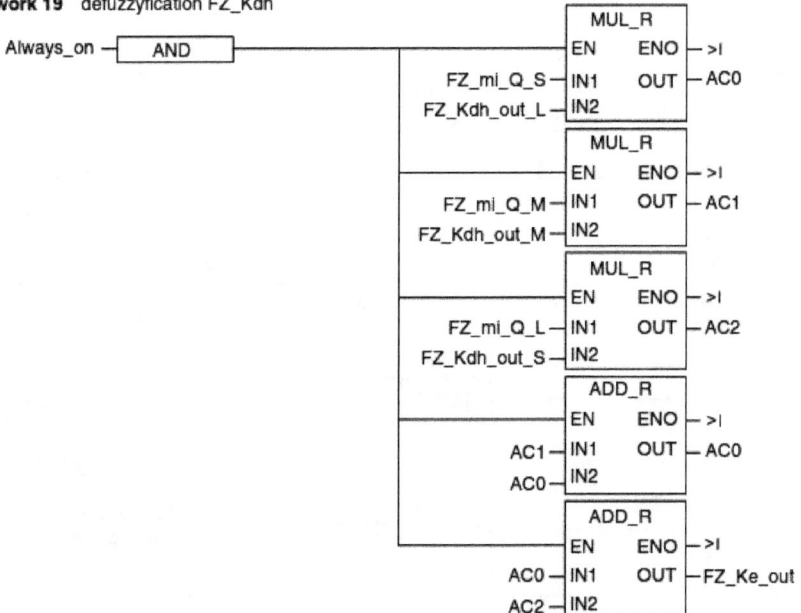

```
Always_on ─┤ AND ├────────────────────┐
                                       │   ┌─────────────────┐
                                       │   │     MUL_R        │
                                       ├───┤EN        ENO├─ >I
                           FZ_mi_Q_S ──┤IN1       OUT├─ AC0
                          FZ_Kdh_out_L ┤IN2           │
                                       │   └─────────────────┘
                                       │   ┌─────────────────┐
                                       │   │     MUL_R        │
                                       ├───┤EN        ENO├─ >I
                           FZ_mi_Q_M ──┤IN1       OUT├─ AC1
                          FZ_Kdh_out_M ┤IN2           │
                                       │   └─────────────────┘
                                       │   ┌─────────────────┐
                                       │   │     MUL_R        │
                                       ├───┤EN        ENO├─ >I
                           FZ_mi_Q_L ──┤IN1       OUT├─ AC2
                          FZ_Kdh_out_S ┤IN2           │
                                       │   └─────────────────┘
                                       │   ┌─────────────────┐
                                       │   │     ADD_R        │
                                       ├───┤EN        ENO├─ >I
                                 AC1 ──┤IN1       OUT├─ AC0
                                 AC0 ──┤IN2           │
                                       │   └─────────────────┘
                                       │   ┌─────────────────┐
                                       │   │     ADD_R        │
                                       └───┤EN        ENO├─ >I
                                 AC0 ──┤IN1       OUT├─ FZ_Ke_out
                                 AC2 ──┤IN2           │
                                           └─────────────────┘
```

7.8.3 PID control of oven temperature

Here we consider the Delta family PLCs. Software for designing Delta family PLCs is WPLSoft. The PID command is used to control the oven temperature when the ambient temperature is unknown. The DVP04PT-S temperature module is used to measure the existing oven temperature and transfer the results to the PLC. The target temperature is 80°C. The PLC executes the automatic parameter setting function (D204 = K3) to apply the best PID parameters and automatically changes the control direction to "Exclusively for the adjusted temperature control" (D204 = K4). The PLC output is stored in D0 as a function result (parameter set), which is the input of the GPWM command. Y0 outputs the PWM pulses (designated by D0) to control the heater, thereby controlling the PID temperature.

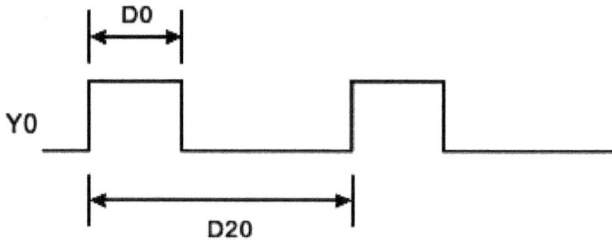

Device	Function
M1	Executing PID instruction
Y0	Outputting adjustable pulses
D0	Storing PID operation result
D10	Storing the target temperature
D11	Storing the present temperature
D20	Storing pulse output cycle of GPWM instruction
D200	Storing PID sampling time parameter

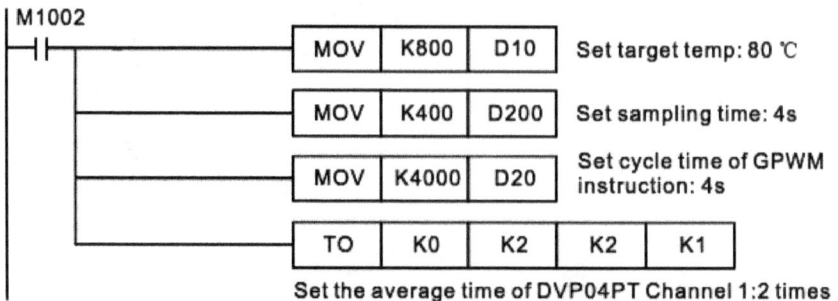

M1002				
	MOV	K800	D10	Set target temp: 80 ℃
	MOV	K400	D200	Set sampling time: 4s
	MOV	K4000	D20	Set cycle time of GPWM instruction: 4s
	TO	K0	K2	K2	K1

Set the average time of DVP04PT Channel 1:2 times

```
M1013
 ┤↑├                          ┌──────┬──────┬──────┬──────┬──────┐
                              │ FROM │  K0  │  K6  │ D11  │  K1  │
                              └──────┴──────┴──────┴──────┴──────┘
                              Sample PV of the oven every 1s and store it in D11
 M0
 ┤ ├                          ┌──────┬──────┬──────┐ Set the control direction
                              │ MOV  │  K3  │ D204 │ as temperature auto-tuning
                              └──────┴──────┴──────┘
                              ┌──────┬──────┐
                              │ RST  │  M0  │
                              └──────┴──────┘
 M1
 ┤ ├                          ┌──────┬──────┬──────┬──────┬──────┐
                              │ PID  │ D10  │ D11  │ D200 │  D0  │
                              └──────┴──────┴──────┴──────┴──────┘
                              Store PID operation result in D200
                              ┌──────┬──────┬──────┬──────┐
                              │ GPWM │  D0  │ D20  │  Y0  │
                              └──────┴──────┴──────┴──────┘
```

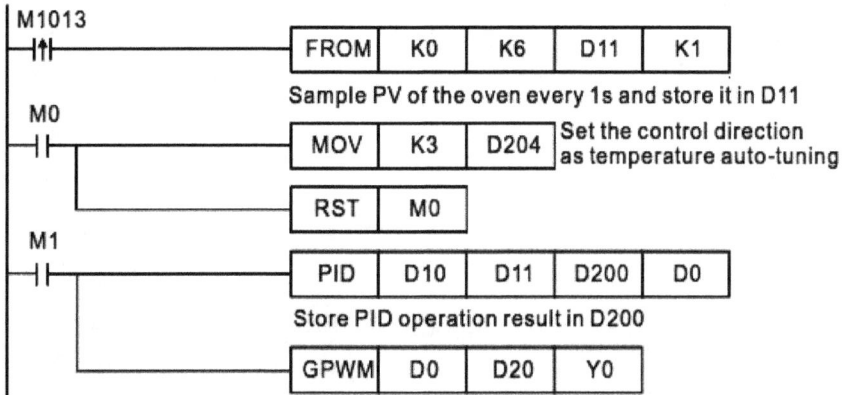

Format of PID instruction:

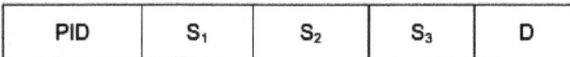

PID	S₁	S₂	S₃	D

$S_1 \rightarrow$ Set value (SV)

$S_2 \rightarrow$ Present value (PV)

$S_3 \rightarrow$ Parameter (Users need to set and adjust it. For the definition, refer to PID parameter
      table in the last part of this example)

$D \rightarrow$ Output value (MV) (D has to be the data register area with latched function)

Device No.	Function	Range	Explanation
(S₃):	Sampling time (T_S) (unit: 10ms)	1~2,000 (unit: 10ms)	If T_S is less than 1 program scan time, PID instruction will be executed for 1 program scan time. If T_S= 0, PID instruction will not be enabled. The minimum T_S has to be longer than the program scan time.
(S₃) +1:	Proportional gain (K_P)	0~30,000 (%)	If SV is bigger than the max. value, the output will be the max. value.
(S₃) +2:	Integral gain (K_I)	0~30,000 (%)	
(S₃) +3:	Differential gain (K_D)	-3,000~30,000 (%)	
(S₃) +4:	Control direction (DIR)		0: automatic control 1: forward control (E = SV - PV) 2: inverse control (E = PV - SV) 3: Auto-tuning of parameter exclusively for the temperature control. The device will automatically become K4 when the auto-tuning is completed and be filled in with the appropriate parameter K_P, K_I and K_D (not available in the 32-bit instruction). 4: Exclusively for the adjusted temperature control (not available in the 32-bit instruction). 5: automatic control(with upper/lower bounds of integral value). Only supported by SV_V1.2 / EH2_V1.2 / SA / SA_V1.8 / SC_V1.6 or higher version PLC.

(S3) +5:	The range that error value (E) doesn't work	0~32,767	Ex: when S3 +5 is set as 5, MV of E between -5 and 5 will be 0.
(S3) +6:	Upper bound of output value (MV)	-32,768~ 32,767	Ex: if S3 +6 is set as 1,000, the output will be 1,000 when MV is bigger than 1,000. S3 +6 has to be bigger or equal S3 +7; otherwise the upper bound and lower bound will switch.
(S3) +7:	Lower bound of output value (MV)	-32,768~ 32,767	Ex: if S3 +7 is set as -1,000, the output will be -1,000 when MV is smaller than -1,000.
(S3) +8:	Upper bound of integral value	-32,768~ 32,767	Ex: if S3 +8 is set as 1,000, the output will be 1,000 when the integral value is bigger than 1,000 and the integration will stop. S3 +8 has to be bigger or equal S3 +9; otherwise the upper bound and lower bound will switch.
(S3) +9:	Lower bound of integral value	-32,768~ 32,767	Ex: if S3 +9 is set as -1,000, the output will be -1,000 when the integral value is smaller than -1,000 and the integration will stop.
(S3) +10,11:	Accumulated integral value	32-bit floating point	The accumulated integral value is only for reference. You can still clear or modify it (in 32-bit floating point) according to your need.
(S3) +12:	The previous PV	-	The previous PV is only for reference. You can still modify it according to your need.
(S3) +13: ↓ **(S3) +19:**	For system use only.		

7.8.4 PID control in SIMATIC S7 TIA Portal

The following steps explain how to execute the above example as a project in the SIMATIC S7-300.

1- Double click (→Totally Integrated Automation Portal V11)

2- In the portal view (→ Create new project → PID → Create)

3- Configure a device (→ First Steps → Configure a device)

4- Add new device (→ Add new device → PLC_1 → CPU317-2 PN/DP → 6ES7 ...→ Add)

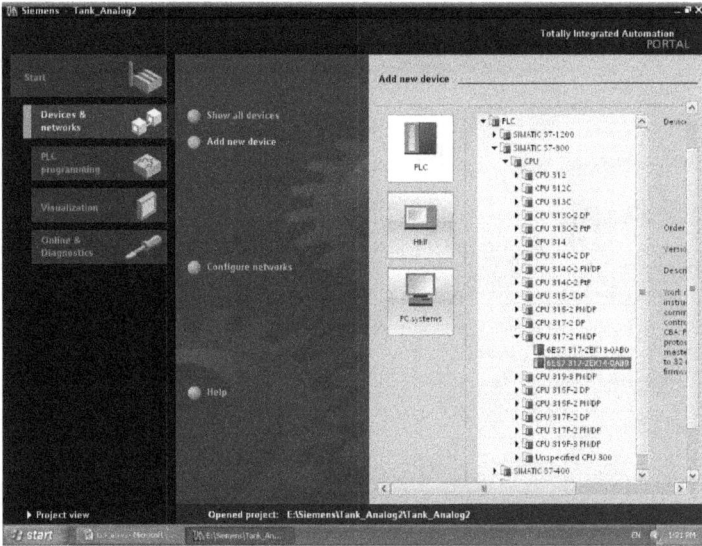

5- With drag & drop we add from the catalog the signal board (→ Catalog → Signal board → AO1 x 12Bit → 6ES7 332-...), (→ Catalog → Signal board → AI1 x 12Bit → 6ES7 331-...), (→ Catalog → Signal board → DI8/DO8×24VDC→ 6ES7 323-...)

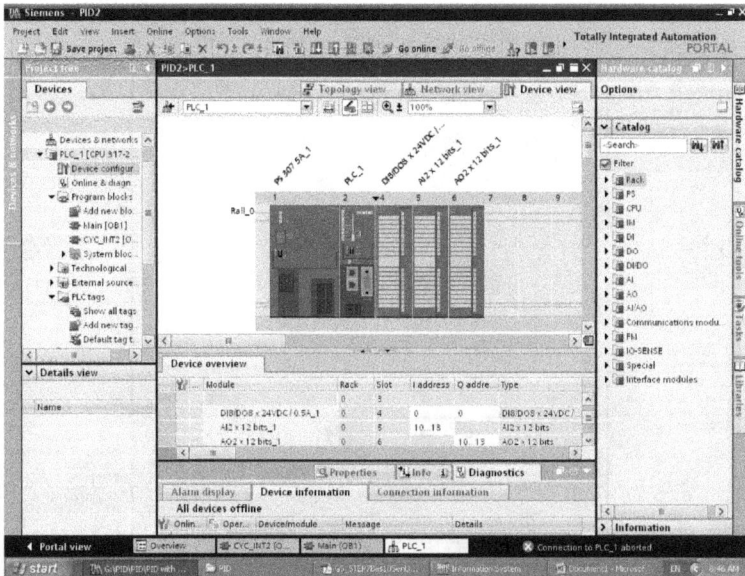

6- Program blocks → Add new block → Time interrupts → Cyclic
   → CYC_INT2[OB32] → OK

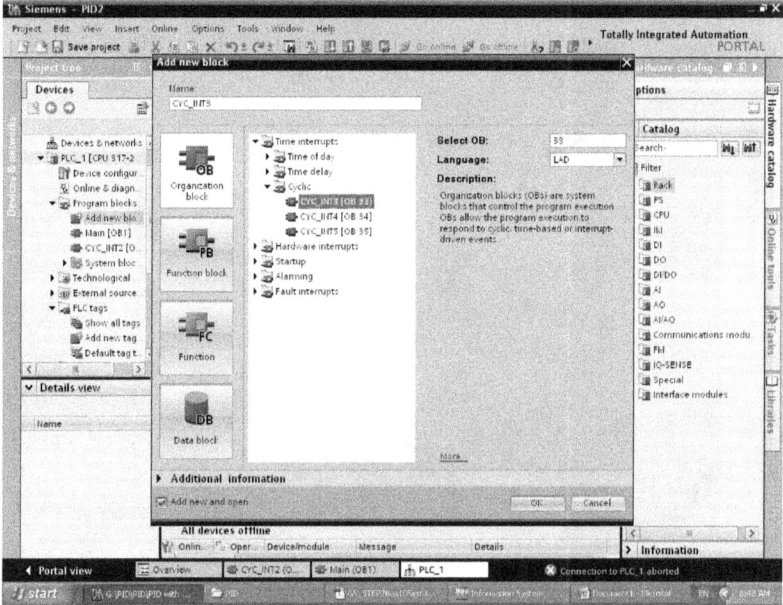

7- PLC tags → Default tags

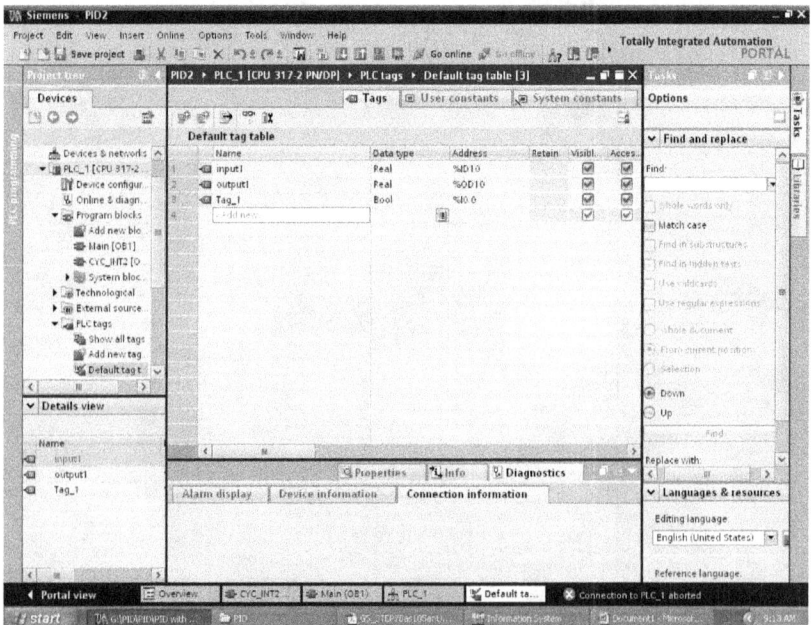

8- Technology → PID Basic functions → Drag and drop CONT_C
   → MAN_ON: Tag_1 → PV_IN: input1 → LMN: output1 →
   SP_INT: 10.0

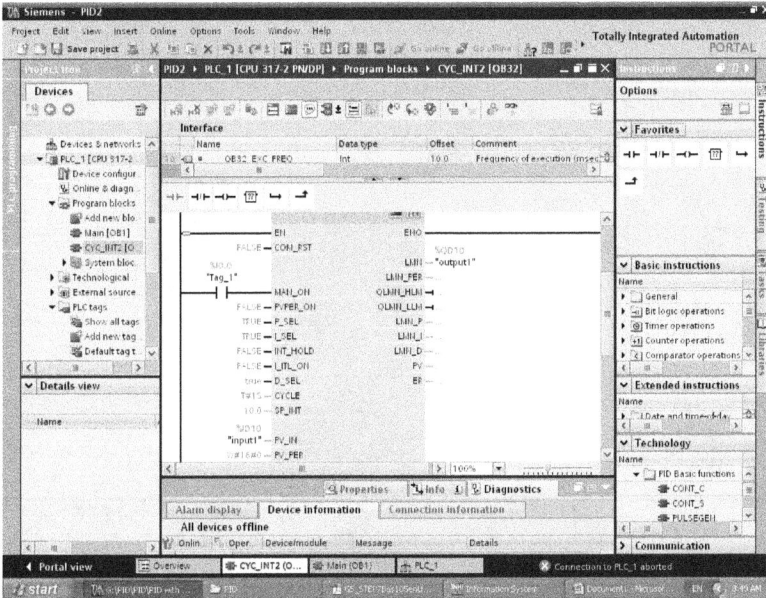

9- Click Configuration window

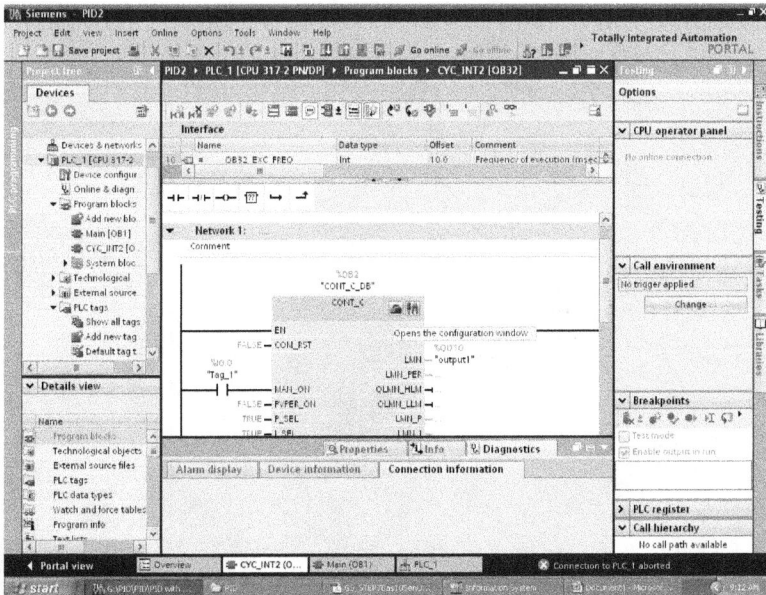

10- Controller → Controller structure → PID

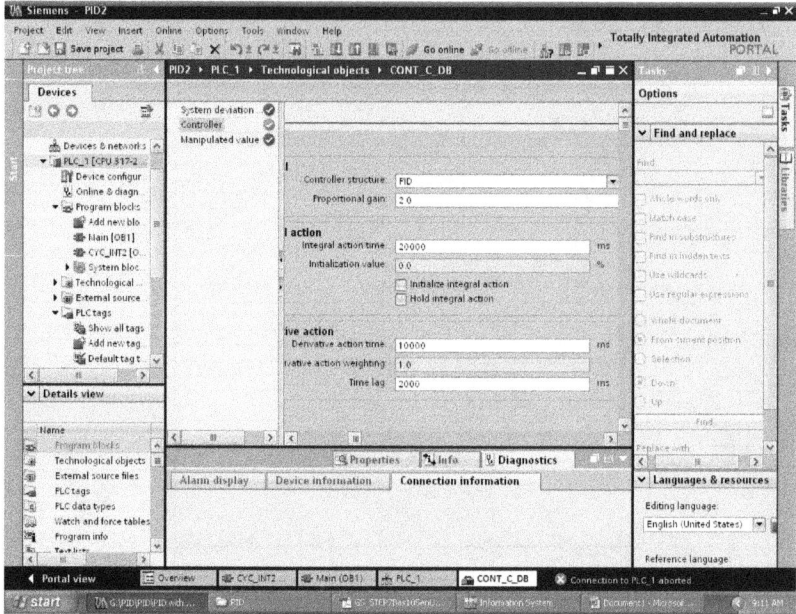

11- PLC_1 → Compile → All

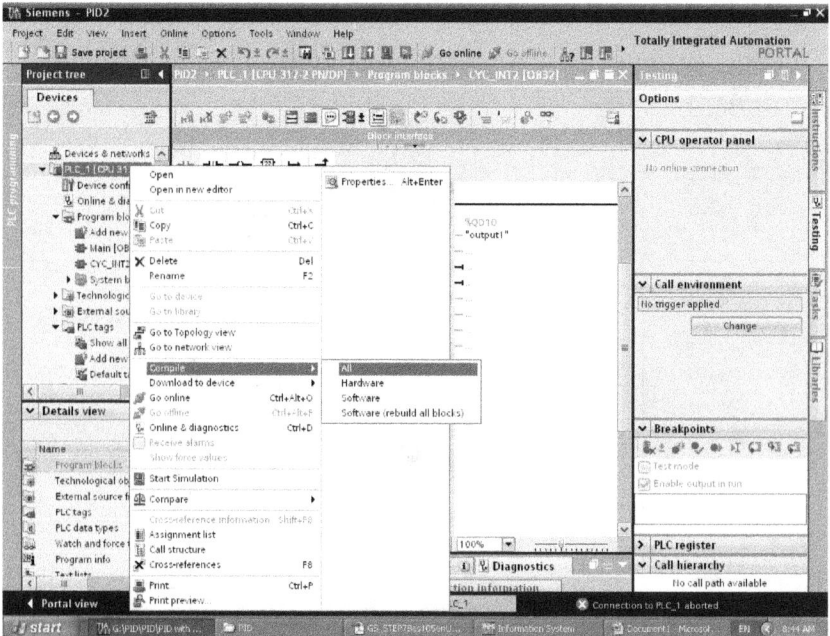

12- PLC_1 → Start Simulation

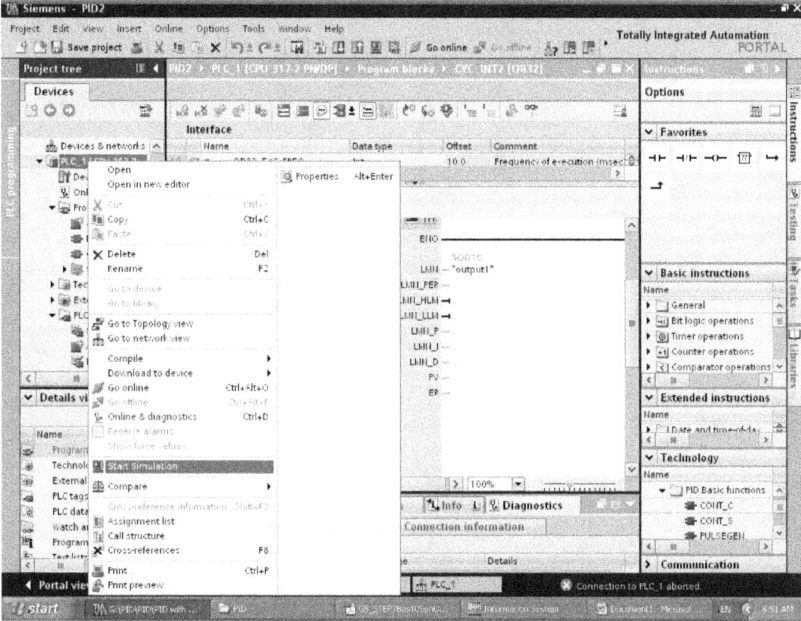

13- Click on 'Load' once more (→ Load).

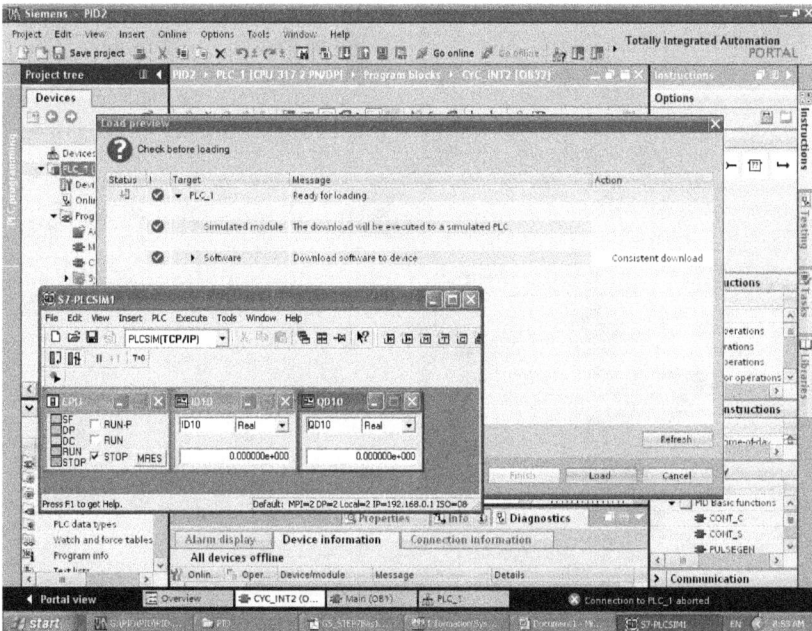

14- Click on the symbol Monitoring on/off → Open commissioning window

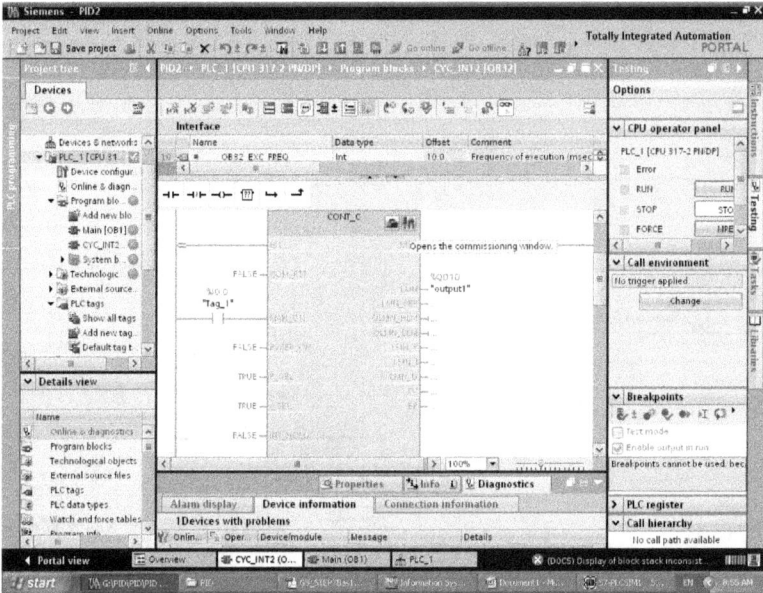

15- Measurement on → RUN-P

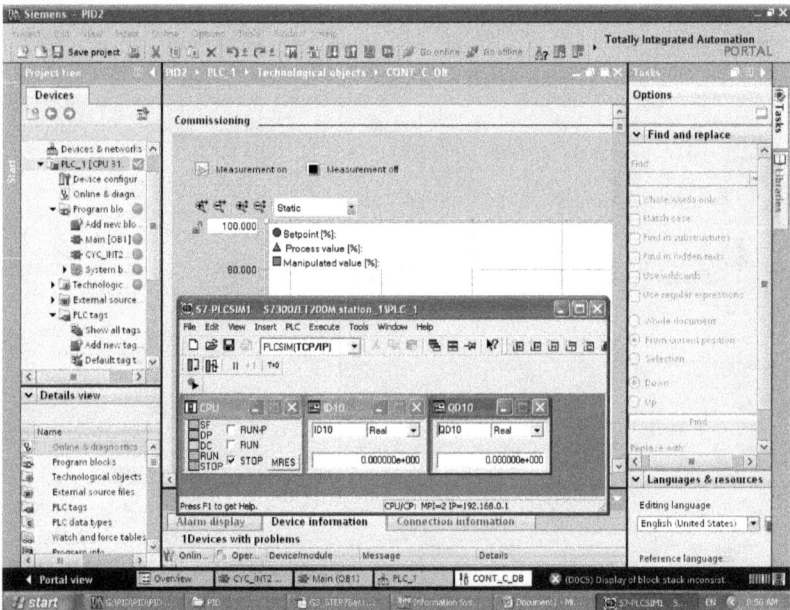

16- With changing input1 (ID10), output1 (QD10) is changing
   and PID curve is sketched.

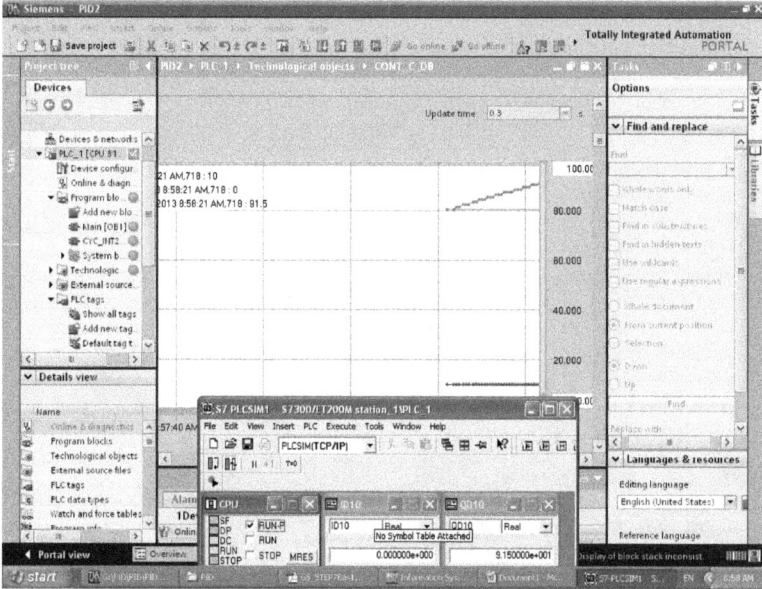

The PID control diagram block is plotted as follows. For automatic
PID control operation, it must be MAN_ON = 0.

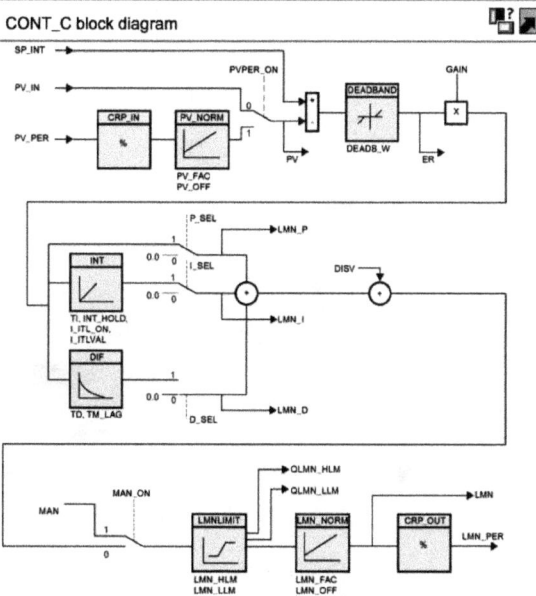

Chapter 8

Application of Ldmicro and PetriLLD Software in Microcontroller Programming

8.1 Introduction

This chapter discusses several control issues using PetriLLD software and Ldmicro software, which are Petri nets and Relay Ladder Logic (RLL) software for the AVR and PIC microcontrollers, respectively. Microcontrollers are much cheaper to use than PLCs, but PLCs are more reliable than microcontrollers are. There are various software available for programming the AVR and PIC microcontrollers, among them the flowcode software, which is a flowchart programming software, can easily generate a hex file for microcontroller programming and can also be simulated well in in its environment. The Proteus software is mainly used to simulate the actual performance of microcontrollers. Using the hex file generated by Ldmicro or Flowcode software and using it in the Proteus software environment, the actual performance of the microcontroller is simulated.

8.2 Introduction to Ldmicro software environment

The Ldmicro software environment shown below has various icons such as File, Edit, Settings, Instruction, Simulate, Compile and Help.

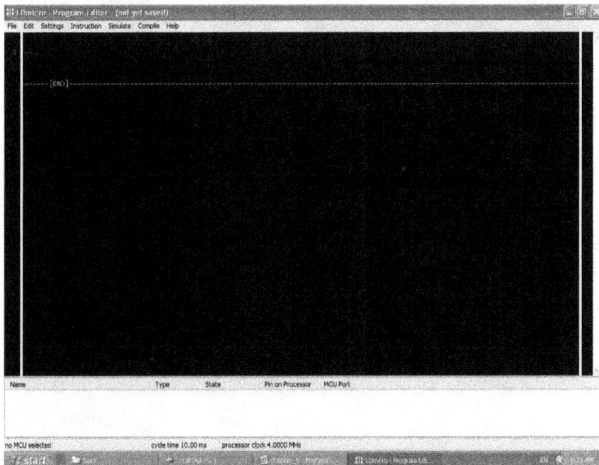

The black window is a relay ladder logic (RLL) environment, and the white window shows the name, type, state, pin on processor, and MCU port including input, output and internal memory signals used in the microcontroller.

The File icon contains the following formats used to create, open, save, and more.

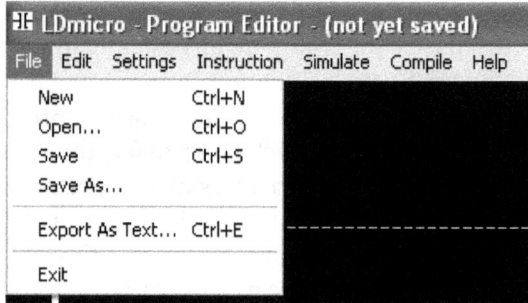

The Edit icon has the following form, which is used to add a rung before and after a rung, delete a selected element, or delete a rung, and so on.

The Settings icon has the following form used to specify the type and parameters of the microcontroller.

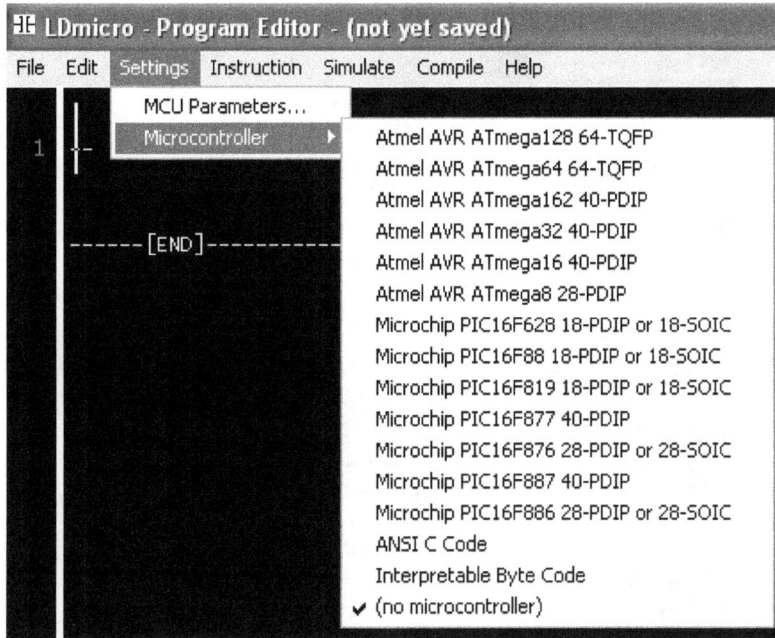

The Instruction icon has the following formats used for relay ladder logic programming (RLL) and includes input contact, output coil, timers, counters, math operations, serial port data transmission, A/D converter, PWM output signal etc.

⚏ LDmicro - Program Editor - (not yet saved)

File   Edit   Settings   Instruction   Simulate   Compile   Help

1 --	Insert Comment	;

Insert Contacts	C

Insert OSR (One Shot Rising)	/
Insert OSF (One Shot Falling)	\

------[END]

Insert TON (Delayed Turn On)	O
Insert TOF (Delayed Turn Off)	F
Insert RTO (Retentive Delayed Turn On)	T

Insert CTU (Count Up)	U
Insert CTD (Count Down)	I
Insert CTC (Count Circular)	J

Insert EQU (Compare for Equals)	=
Insert NEQ (Compare for Not Equals)	
Insert GRT (Compare for Greater Than)	>
Insert GEQ (Compare for Greater Than or Equal)	.
Insert LES (Compare for Less Than)	<
Insert LEQ (Compare for Less Than or Equal)	,

Insert Open-Circuit
Insert Short-Circuit
Insert Master Control Relay

Insert Coil	L
Insert RES (Counter/RTO Reset)	E

Insert MOV (Move)	M
Insert ADD (16-bit Integer Add)	+
Insert SUB (16-bit Integer Subtract)	-
Insert MUL (16-bit Integer Multiply)	*
Insert DIV (16-bit Integer Divide)	D

Insert Shift Register
Insert Look-Up Table
Insert Piecewise Linear
Insert Formatted String Over UART

Insert UART Send
Insert UART Receive
Insert Set PWM Output

Insert A/D Converter Read	P

Insert Make Persistent

Make Normal	A
Make Negated	N
Make Set-Only	S
Make Reset-Only	R

The Simulate icon has the following formats that used to simulate relay ladder logic (RLL) programming.

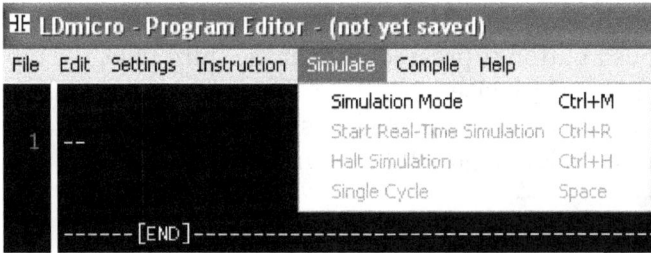

The Compile icon has the following formats used to compile and generate a hex file.

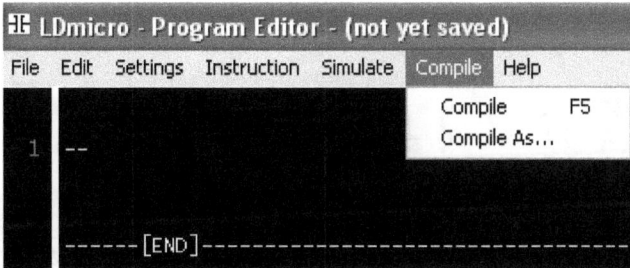

The Help icon has the following form, which is the Ldmicro software user guide.

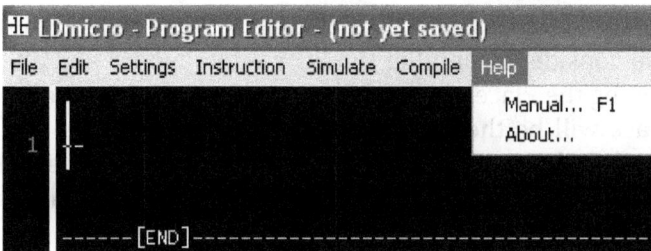

8.3 Performing an automation problem with Ldmicro and PetriLLD

We want to control a motor with just a push button as follows:

- When the push-button is pressed once, the motor will run clockwise.
- When the push button is pressed once, while the motor is rotating, the motor stops.
- If the push button is pressed once at the third time, the motor will rotate counterclockwise.

The Petri net for this problem is illustrated below. The order of writing the relay ladder logic (RLL) program is as follows:

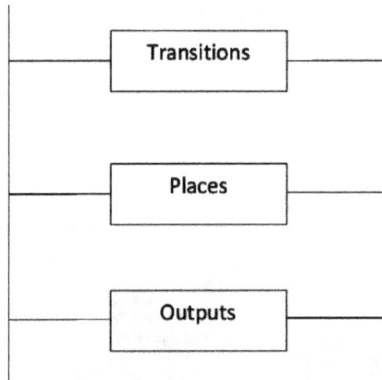

- Writing the Transitions program is such that all the places (in parallel) and inputs (in series) leading to the Transition are considered together and the Transition will be the corresponding RLL rung output. The Transitions program does not include parallel NO contact of the related transition coil.
- Writing the Places program is such that all Transitions ending in the Place are considered parallel (in NO contact mode) and all Transitions leaving the Place are considered series NC contacts and the desired place will be the corresponding RLL step output. The Places program includes parallel NO contact of the related place coil. Writing the idle (Primary Place) RLL step is a bit different, and its program includes NC contact for other places in series.
- Writing the Outputs program is exactly like the Places program.
Using the Atmel AVR ATmega32 40-PDIP microcontroller, the Transitions program is as follows.

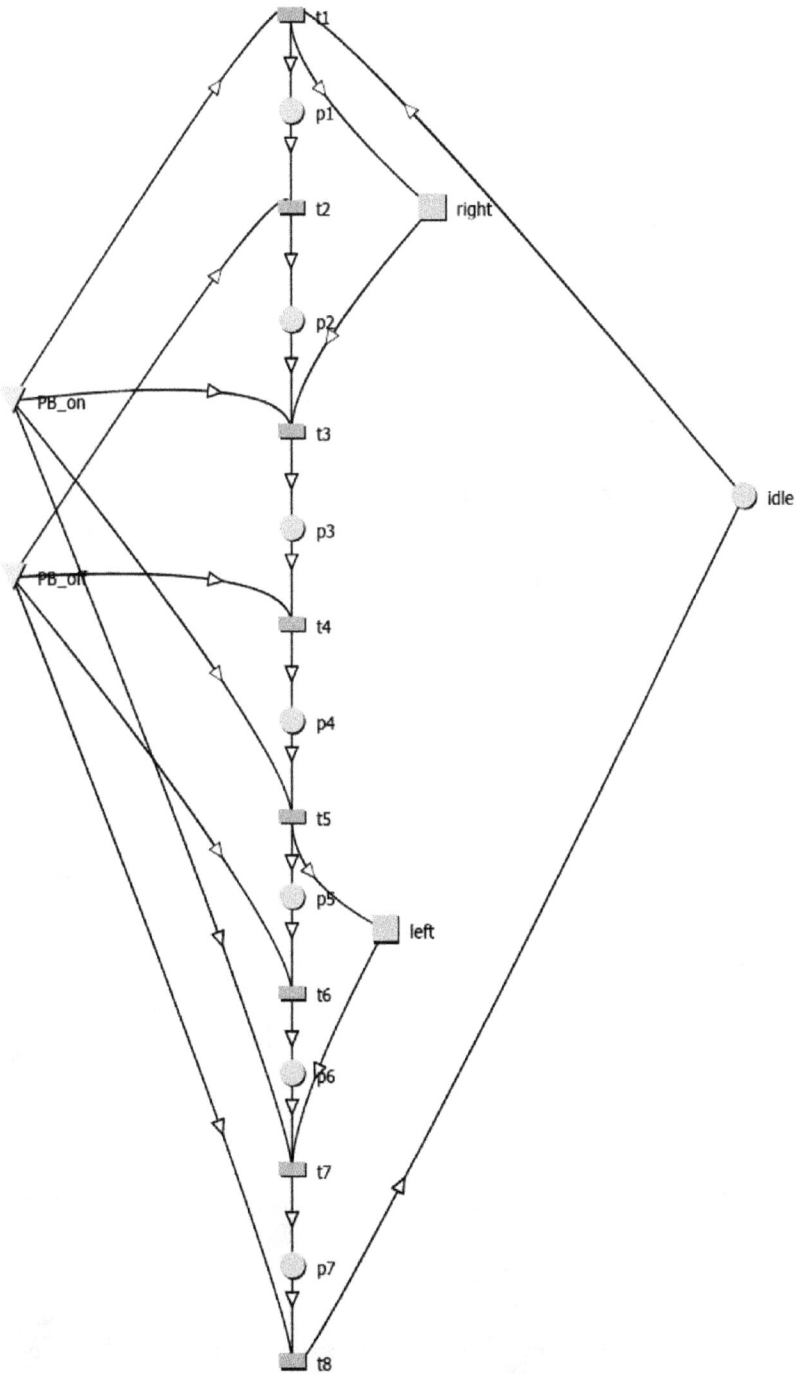

```
LDmicro export text
for 'Atmel AVR ATmega32 40-PDIP', 4.000000 MHz crystal, 10.0 ms cycle time

LADDER DIAGRAM:
```

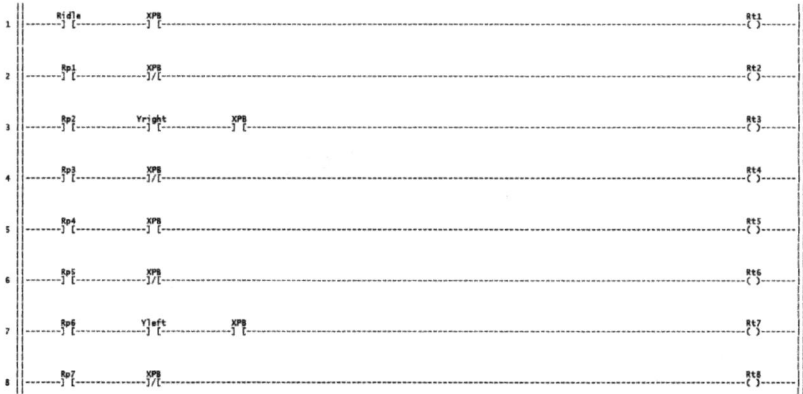

The Places program is as follows.

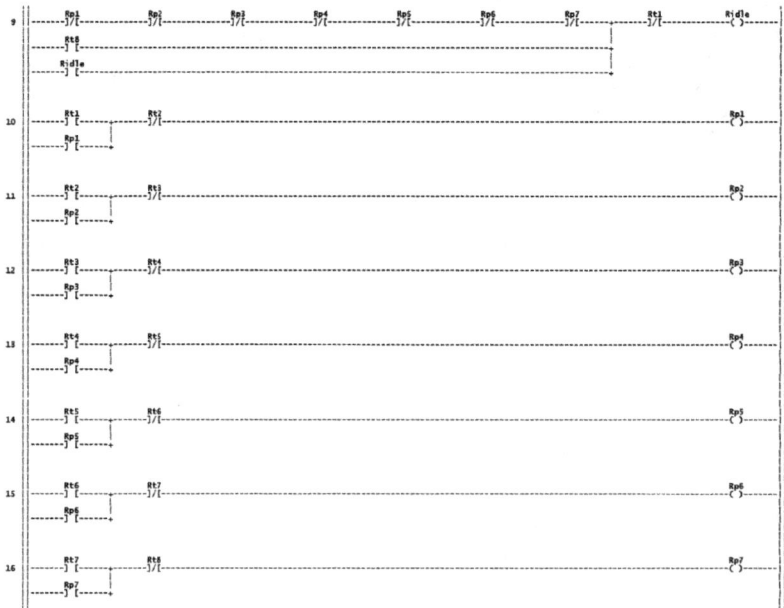

The Outputs program (exactly like the Places program) is as follows.

The digital input and output signals and the internal relay and their dedicated microcontroller pins are shown below.

```
I/O ASSIGNMENT:

    Name                        | Type            | Pin
    ----------------------------+-----------------+------
    XPB                         | digital in      | 22
    Yleft                       | digital out     | 24
    Yright                      | digital out     | 26
    Ridle                       | int. relay      |
    Rp1                         | int. relay      |
    Rp2                         | int. relay      |
    Rp3                         | int. relay      |
    Rp4                         | int. relay      |
    Rp5                         | int. relay      |
    Rp6                         | int. relay      |
    Rp7                         | int. relay      |
    Rt1                         | int. relay      |
    Rt2                         | int. relay      |
    Rt3                         | int. relay      |
    Rt4                         | int. relay      |
    Rt5                         | int. relay      |
    Rt6                         | int. relay      |
    Rt7                         | int. relay      |
    Rt8                         | int. relay      |
```

By compiling the program, generated hex file can be used for microcontroller programming. Using the Proteus software, you can do the actual simulation using the generated hex file shown below.

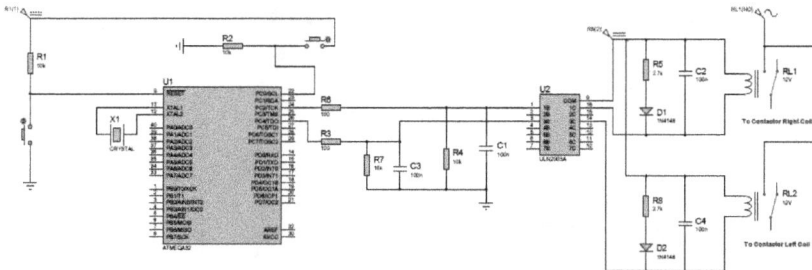

Since the microcontroller output signal has low voltage and current and cannot drive the 12V relay, a driver (ULN2003A) is used as shown in the figure between the microcontroller and the 12V relay. Each relay is connected to the coil of the clockwise and counterclockwise contactor. Therefore, at a low cost, this automation problem can be implemented with a microcontroller instead of a PLC.

8.4 Another automation problem (Counter and Timer Combination)

The performance order of the system shown in the following figure is:
- The box is in LS1 position (LS1 closed).
- By pressing the start button, the conveyor motor starts and the box moves to position A (LS1 opens).
- The conveyor moves the box to position A and then stops (the position is detected by eight off-to-on pulses from the encoder to up counter).
- After a 10 second delay, the conveyor starts to move and the box moves to LS2 and stops (LS2 closes).
- An emergency stop push button is used to stop the process at any time.
- If the process is stopped by the emergency stop push button, the timer and counter will reset automatically.

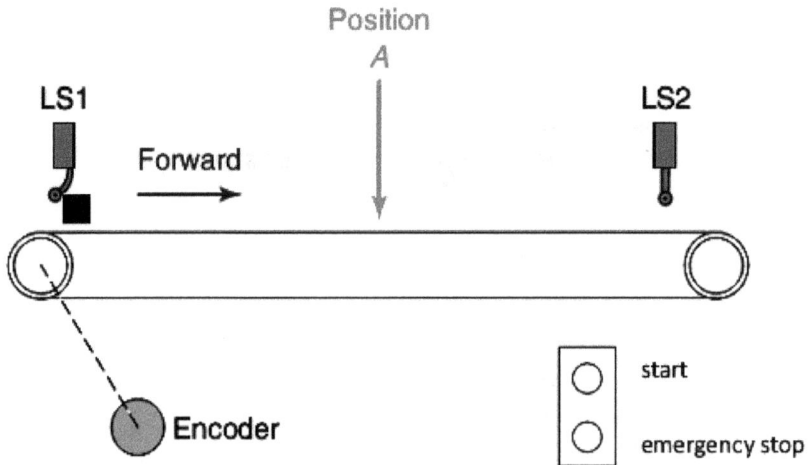

The Petri net for this problem is illustrated below. The order of writing the relay ladder logic (RLL) program is as follows:

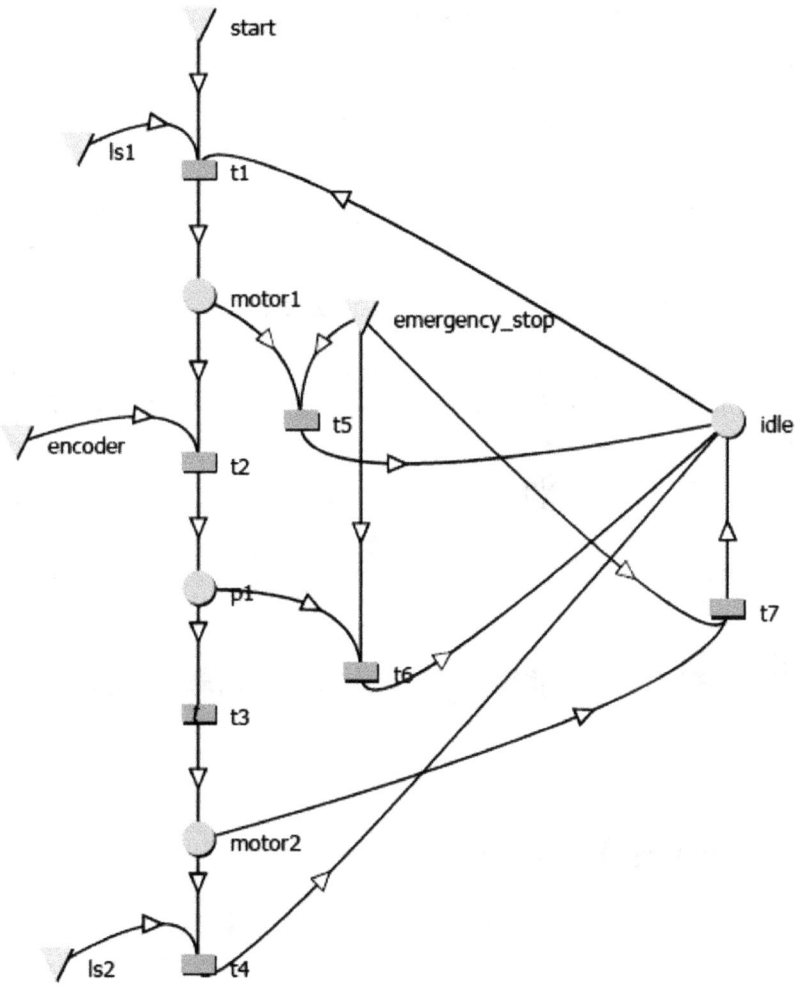

Using the Atmel AVR ATmega32 40-PDIP microcontroller, the Transitions program is as follows.

```
LDmicro export text
for 'Atmel AVR ATmega32 40-PDIP', 4.000000 MHz crystal, 10.0 ms cycle time
```

LADDER DIAGRAM:

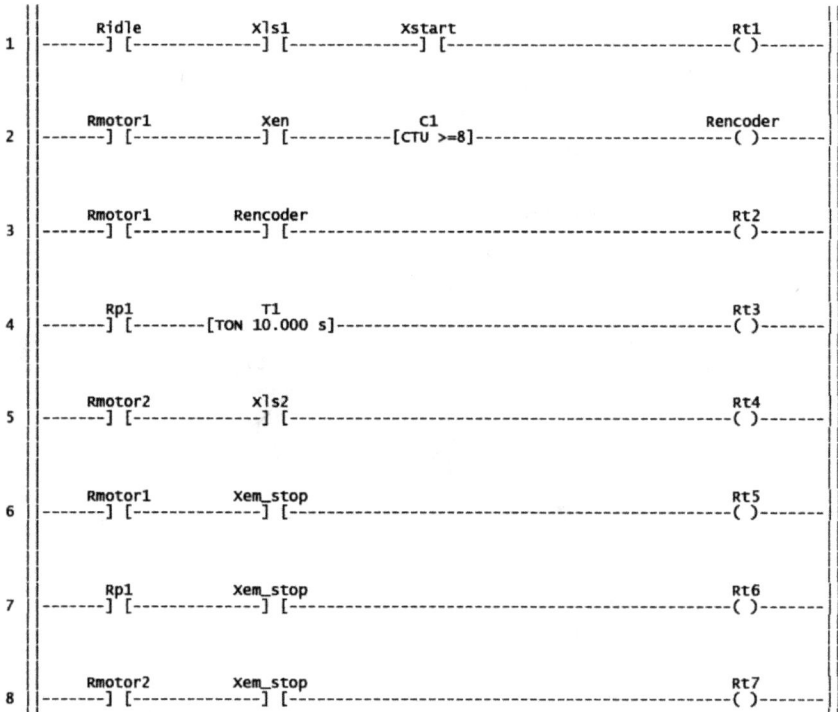

```
      ||   Ridle            Xls1            Xstart                      Rt1      ||
  1   ||------] [-------------] [-------------] [-----------------------( )------||
      ||                                                                        ||
      ||                                                                        ||
      ||   Rmotor1           Xen             C1                        Rencoder ||
  2   ||------] [-------------] [----------[CTU >=8]--------------------( )------||
      ||                                                                        ||
      ||                                                                        ||
      ||   Rmotor1           Rencoder                                  Rt2      ||
  3   ||------] [-------------] [----------------------------------------( )------||
      ||                                                                        ||
      ||                                                                        ||
      ||   Rp1              T1                                         Rt3      ||
  4   ||------] [--------[TON 10.000 s]----------------------------------( )------||
      ||                                                                        ||
      ||                                                                        ||
      ||   Rmotor2           Xls2                                      Rt4      ||
  5   ||------] [-------------] [----------------------------------------( )------||
      ||                                                                        ||
      ||                                                                        ||
      ||   Rmotor1           Xem_stop                                  Rt5      ||
  6   ||------] [-------------] [----------------------------------------( )------||
      ||                                                                        ||
      ||                                                                        ||
      ||   Rp1              Xem_stop                                   Rt6      ||
  7   ||------] [-------------] [----------------------------------------( )------||
      ||                                                                        ||
      ||                                                                        ||
      ||   Rmotor2           Xem_stop                                  Rt7      ||
  8   ||------] [-------------] [----------------------------------------( )------||
```

The Places program is as follows.

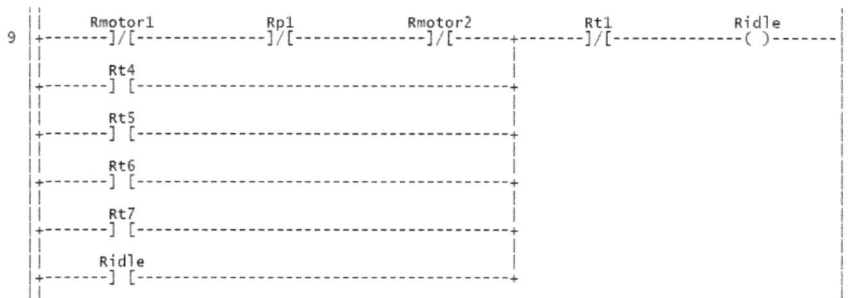

```
      ||   Rmotor1           Rp1            Rmotor2        Rt1          Ridle    ||
  9   +----]/[-------------]/[-------------]/[------+------]/[-------------( )------||
      ||   Rt4                                      |                             ||
      +------] [----------------------------------+                             ||
      ||   Rt5                                      |                             ||
      +------] [----------------------------------+                             ||
      ||   Rt6                                      |                             ||
      +------] [----------------------------------+                             ||
      ||   Rt7                                      |                             ||
      +------] [----------------------------------+                             ||
      ||   Ridle                                    |                             ||
      +------] [----------------------------------+                             ||
```

```
           Rt1              Rt2              Rt5                          Rmotor1
10   ||-------] [------+-------]/[-------------]/[----------------------------( )-------||
           Rmotor1        |
     ||-------] [------+



           Rt2              Rt3              Rt6                            Rp1
11   ||-------] [------+-------]/[-------------]/[----------------------------( )-------||
           Rp1            |
     ||-------] [------+



           Rt3              Rt4              Rt7                          Rmotor2
12   ||-------] [------+-------]/[-------------]/[----------------------------( )-------||
           Rmotor2        |
     ||-------] [------+



           Rt4                                                             C1
13   ||-------] [------+-----------------------------------------------------{RES}------||
           Xem_stop       |
     ||-------] [------+



           Rt4                                                             T1
14   ||-------] [------+-----------------------------------------------------{RES}------||
           Xem_stop       |
     ||-------] [------+
```

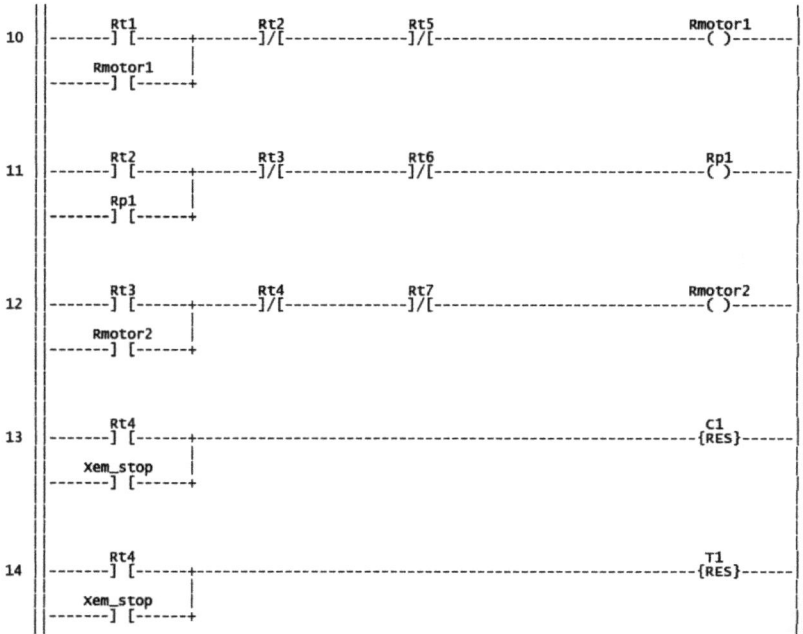

The Outputs program is as follows:

```
           Rmotor1                                                       Ymotor
15   ||-------] [------+-----------------------------------------------------( )-------||
           Rmotor2        |
     ||-------] [------+


     ||------[END]---------------------------------------------------------------------||
```

The digital input and output signals and the internal relay and their dedicated microcontroller pins are shown below.

```
I/O ASSIGNMENT:
```

Name	Type	Pin
Xem_stop	digital in	22
Xen	digital in	24
Xls1	digital in	26
Xls2	digital in	28
Xstart	digital in	1
Ymotor	digital out	3
Rencoder	int. relay	
Ridle	int. relay	
Rmotor1	int. relay	
Rmotor2	int. relay	
Rp1	int. relay	
Rt1	int. relay	
Rt2	int. relay	
Rt3	int. relay	
Rt4	int. relay	
Rt5	int. relay	
Rt6	int. relay	
Rt7	int. relay	
T1	turn-on delay	
C1	counter	

By compiling the program, generated hex file can be used for micro-controller programming. Using the Proteus software, you can do the actual simulation using the hex file as shown below.

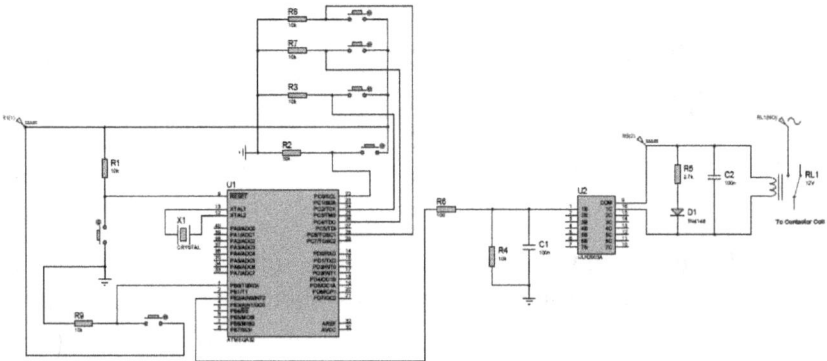

Since the microcontroller output signal has low voltage and current and cannot drive the 12V relay, a driver (ULN2003A) is used as shown in the figure between the microcontroller and the 12V relay. The relay output NO contact is connected to the coil of the contactor. Therefore, at a low cost, this automation problem can be implemented with a microcontroller instead of a PLC.

8.5 AVR ATMEGA32 Based PLC Hardware

The AVR ATMEGA32 based PLC is implemented to get experimental results. A photograph of the hardware implementation is shown as follows:

The circuit schematic and PCB design of the AVR ATMEGA32 based PLC have been accomplished using Proteus software. The PLC hardware includes a mainboard, CPU (AVR ATMEGA32 microcontroller) and the programmer. The designed circuit schematic and PCB of each part are depicted in the following pictures.

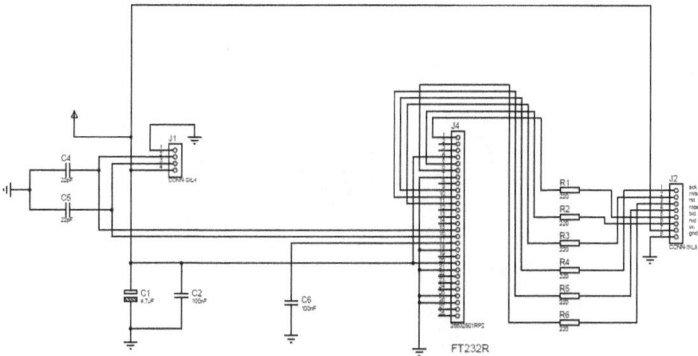

FT232R Synchronous BitBang (diecimila) Programmer Schematic

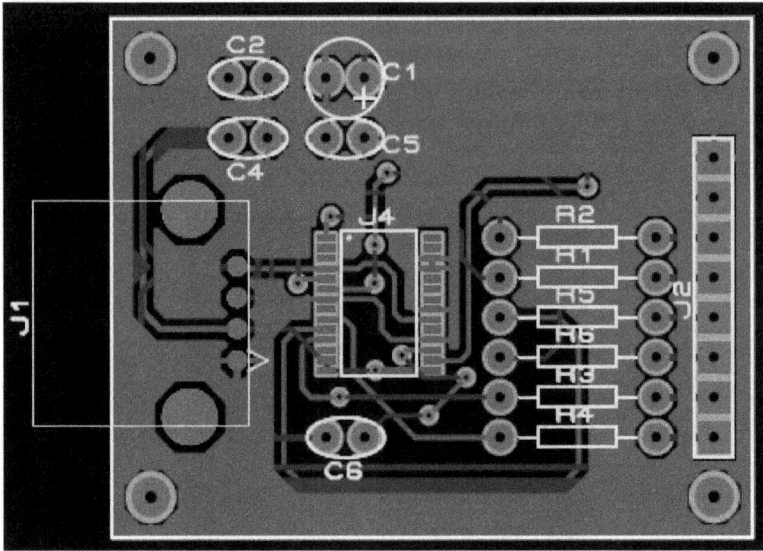

FT232R Synchronous BitBang (diecimila) Programmer PCB

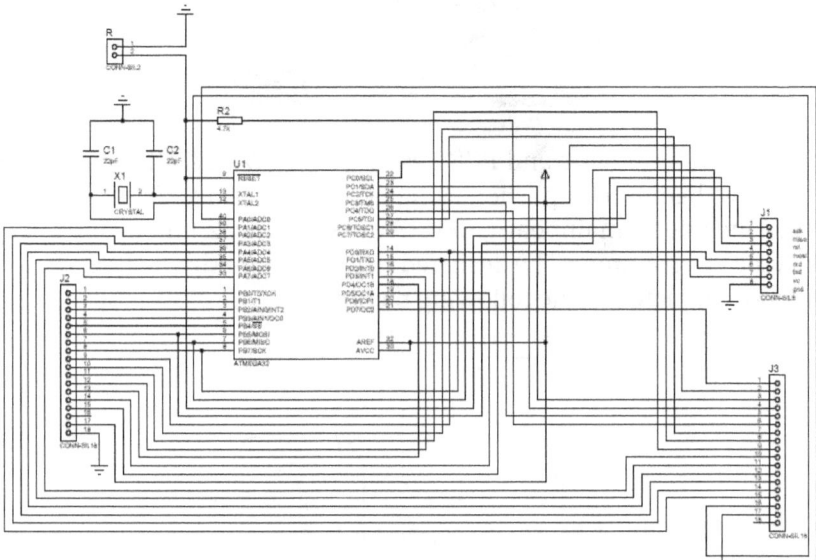

CPU (AVR ATMEGA32) Schematic

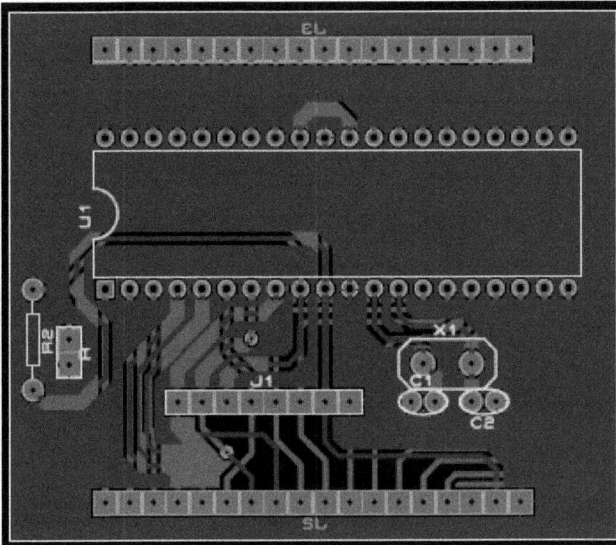

CPU (AVR ATMEGA32) PCB

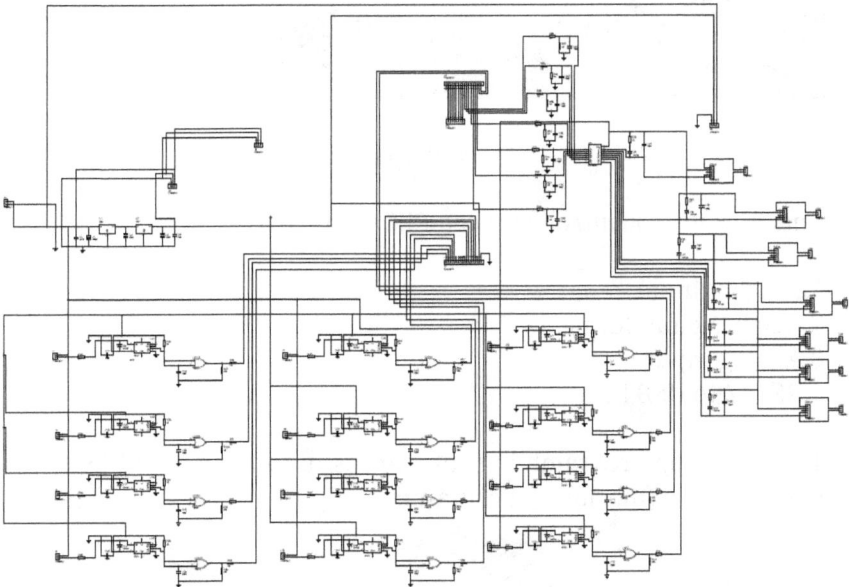

PLC Mainboard Schematic

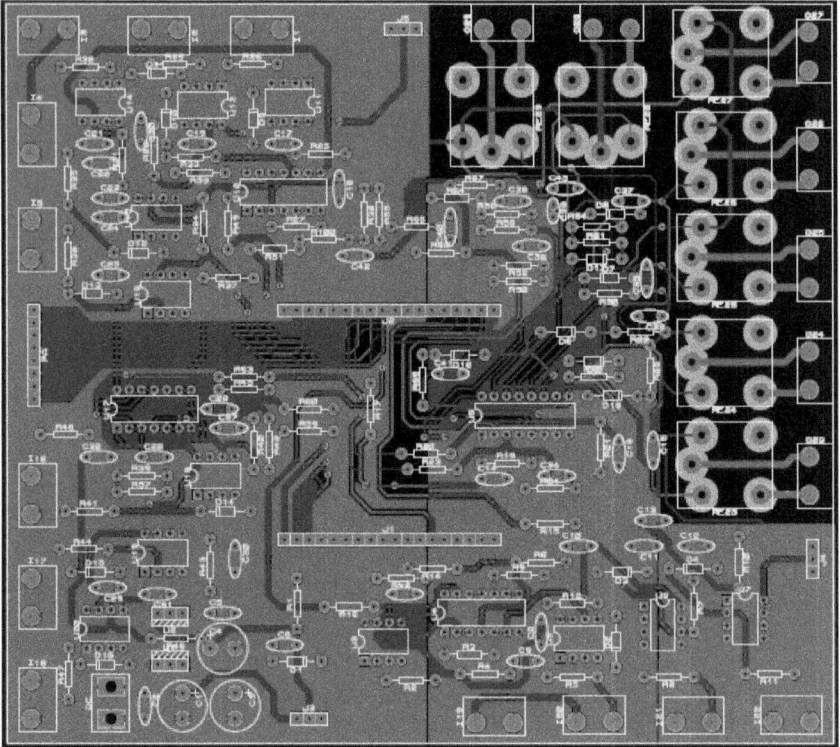

PLC Mainboard PCB

This AVR microcontroller based PLC includes:

12 Digital Input
7 Relay Digital Output
8 Analog Input
1 RS232 to USB Port

The following programming software is used for FT232R Synchronous BitBang (diecimila) Programmer.

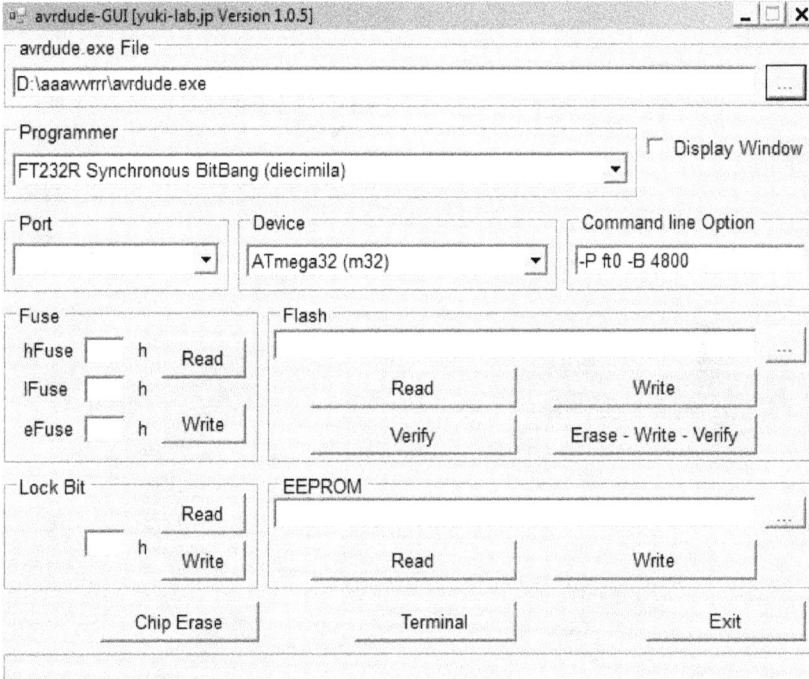

Programming software used for FT232R Synchronous BitBang
(diecimila) Programmer

Chapter 9

Low Cost Arduino Based PLCs

9.1 Introduction

Arduino is an open-source electronics platform based on simple hardware and software environment. Arduino boards are able to read inputs for examples light on a sensor, pressing a button, or a message - and turn it into an output - activating a motor, turning on an LED, etc. Arduino boards could do what to do by programming the microcontroller on the board. To do so you use the Arduino programming language (based on Wiring), and the Arduino Software (IDE), based on Processing. Arduino has been used in thousands of projects, from everyday objects to complex scientific instruments. Arduino is an open-source platform and its users have added up to an incredible amount of knowledge that can be of great help to experts. Arduino was born at the Ivrea Interaction Design Institute as an easy tool for fast prototyping, aimed at users without a background in electronics and programming. As soon as it was applied extensively worldwide, the Arduino board started changing to adapt to new needs and new challenges. All Arduino boards are completely open-source, helping users to build experimental boards independently and eventually adapt them to their particular needs. The software is also open-source, and it is growing through the vast users worldwide. Low cost Arduino boards can be used instead of PLC for industrial automation purposes. PlcLib library presented by Walter Ditch for Arduino could be applied for sequential function chart (SFC) programming. Moreover, Ldmicro software is also used for relay ladder logic (RLL) programming of Arduino boards. In this chapter, an industrial automation example with an Arduino board is evaluated with both SFC and RLL programming methods.

9.2 PlcLib library for Arduino

The order of performance of the system shown in the following figure is:

- The box is in LS1 position (LS1 closed).

- By pressing the start button, the conveyor motor starts and the box moves to position A (LS1 opens).
- The conveyor moves the box to position A and then stops (the position is detected by eight off-to-on pulses from the encoder to up counter).
- After a 10 second delay, the conveyor starts to move and the box moves to LS2 and stops (LS2 closes).
- An emergency stop push button is used to stop the process at any time.
- If the process is stopped by the emergency stop push button, the timer and counter will reset automatically.

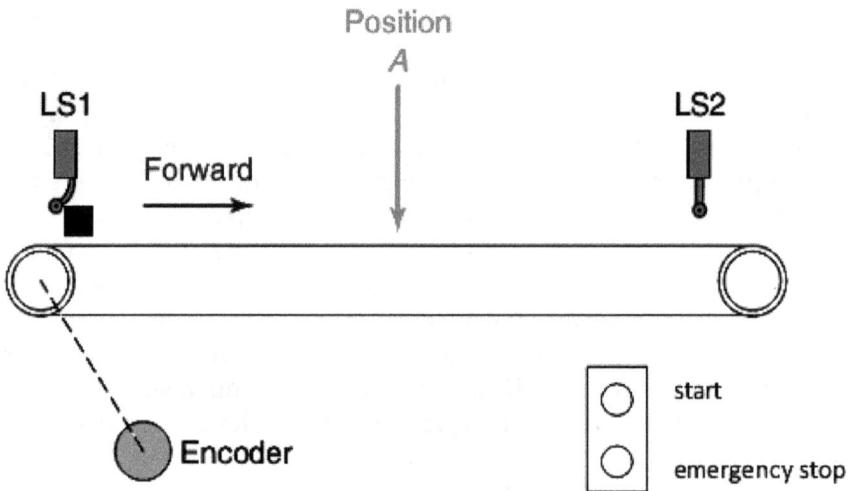

The Petri net for this problem is illustrated below.

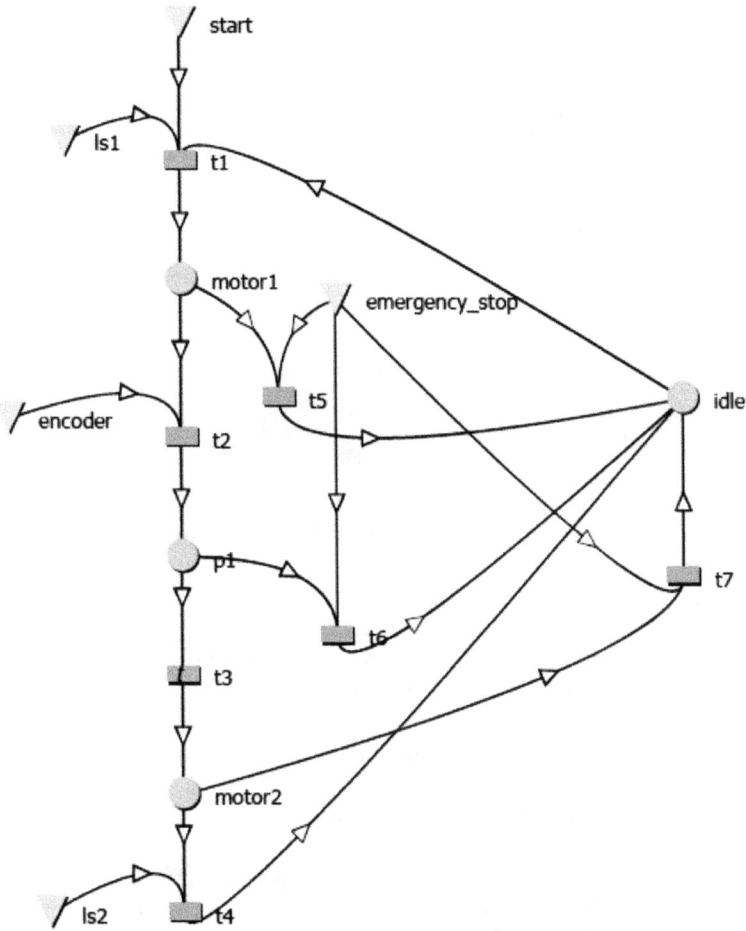

The following code is used in Arduino IDE environment to implement above automation system with SFC programming. The following default pins are defined in PlcLib library. However, we need 5 input and one output so we use CustomIO in library examples to define pins as necessary in this automation control example. The CustomIo.ino file is as follows.

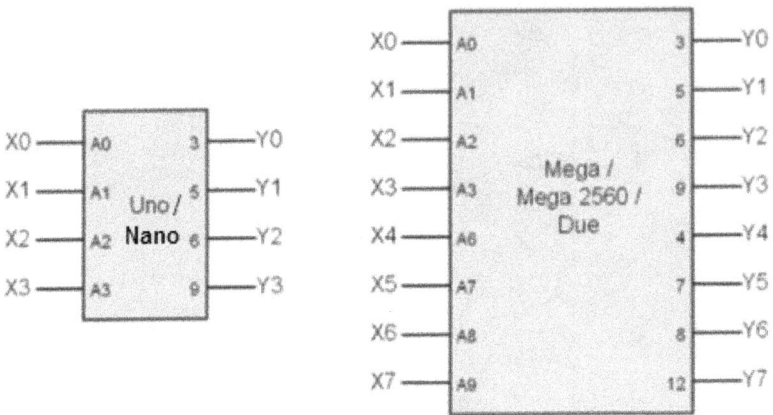

```
#define noPinDefs
#include <plcLib.h>
/* Programmable Logic Controller Library for the Arduino and Compatibles
   Custom I/O - Define your own inputs and outputs from scratch
   Begin by including the line '#define noPinDefs' at the start, which prevents
   creation of standard inputs and output names X0, X1, ..., Y0, Y1, ... etc.
   (This option is available in plcLib V1.2.0 or greater.)
   The 'setupPLC();' command is not needed if you are using custom inputs
   and outputs.
   You can then define your own pin names, as required. Any 'unsigned inte-
ger'
   variables are treated as user defined variables, while 'integers' refer to
   local pin numbers. For example: -
      "unsigned int X1, Y1;" creates X1 and Y1 as user defined variables
      "int X0=A0, Y0=3, Y2=6;" creates X0, Y0 and Y2 as local pin names
   The next step is to define pins as either inputs or outputs by using the
   "pinMode(pin, mode)" command. Sample configuration code has been
placed in
   the 'customIO' function which is available from the IO tab and called from
   within the 'setup()' function below.
   To enable remote monitoring of inputs and outputs, begin by enabling the
   serial port in the setup(); section. E.g. "Serial.begin(9600);"
      Next, include the serial monitor command in the loop() section. E.g.
      "serialMonitor("YourCircuitBoardNameHere");"
   Software and Documentation:
   https://github.com/wditch/plcLib
*/
unsigned int Idle = 1;      // Create variables
unsigned int motor1 = 0;
unsigned int p1 = 0;
unsigned int motor2 = 0;
```

```
unsigned int cout = 0;
int X0=A0, X1=A1, X2=A2, X3=A3, X4=A4, Y0=3;    // Create local pins
//X0=start,  X1=ls1,  X2=ls2,  X3=encoder,  X4=emergency_stop,  Y0=mo-
tor1//motor2
Counter ctr(8); // Final count = 8, starting at zero
//unsigned long TIMER0 = 0; // Define variable used to hold timer 0
elapsed time
unsigned long DELAY0 = 0;
void setup() {
 customIO();          // Setup input and output pin directions (See IO tab)
 // Serial.begin(9600); // Enable serial port (needed for serial IO monitor)
}
void loop() {
 in(Idle); // Read Start-up state
 andBit(X0); // AND with Step 1 transition input
 andBit(X1);
 set(motor1); // Activate Step 1
 reset(Idle); // Cancel Start-up state
 in(motor1);
 andBit(X4);
 ctr.clear();
 set(Idle);
 reset(motor1);
 in(motor1);
 andBit(X3);
 //timerOn(TIMER0, 10); // 10 ms switch debounce delay
 ctr.countUp(); // Count up
 ctr.upperQ(); // Display Count Up output on Y1
 out(cout);
 andBit(cout); // Read Input 0
 set(p1); // Activate Step 1
 reset(motor1); // Cancel Start-up state
 in(p1);
 andBit(X4);
 ctr.clear();
 set(Idle);
 reset(p1);
 in(p1);
 timerOn(DELAY0, 10000);
 set(motor2); // Activate Step 1
 reset(p1); // C

 in(motor2);
 andBit(X4);
 ctr.clear();
```

```
  set(Idle);
  reset(motor2);
  in(motor2);
  andBit(X2);
  ctr.clear();
  set(Idle);
  reset(motor2);
  in(motor1);
  orBit(motor2);
  out(Y0);
  //serialMonitor("YourCircuitBoardNameHere");    // Enable remote I/O
monitoring via the serial port
}
```

The IO.ino file is given below.

```
void customIO() {
  // Input pin directions
  pinMode(X0, INPUT);
  pinMode(X1, INPUT);
  pinMode(X2, INPUT);
  pinMode(X3, INPUT);
  pinMode(X4, INPUT);
  // Output pin directions
  pinMode(Y0, OUTPUT);
  // Note that Xs and Ys are variables so
  // no pinMode definitions are required
}
```

The simulation with Arduino Nano board using Proteus software and hex file produced using Arduino IDE gives the expected result as follows.

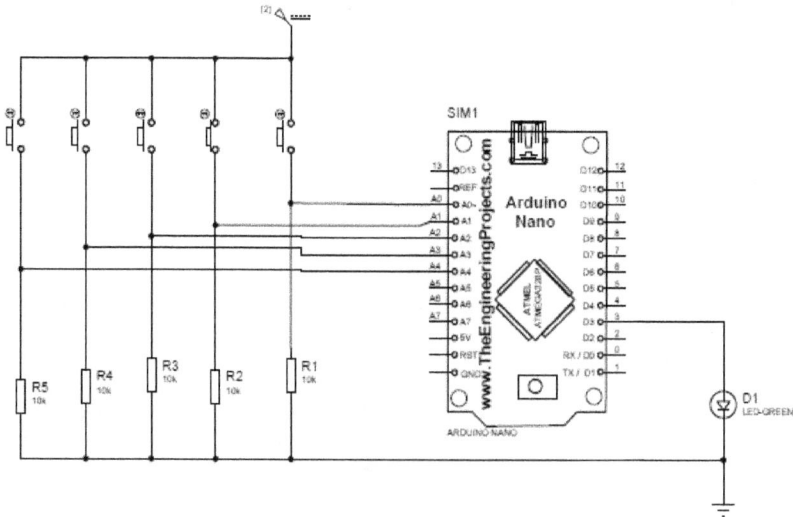

9.3 Arduino RLL programming using Ldmicro software

We consider the previous example. The ladder logic (RLL) program is written as described before. The Transitions program is as follows.

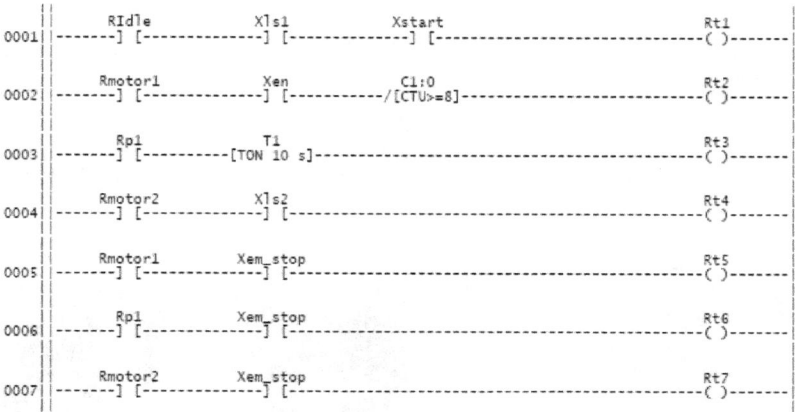

```
          ||  RIdle          X1s1            Xstart                      Rt1
0001 ||------] [-------------] [-------------] [----------------------( )------||
          ||
          ||  Rmotor1         Xen            C1:0                        Rt2
0002 ||------] [-------------] [---------/[CTU>=8]---------------------( )------||
          ||
          ||  Rp1             T1                                        Rt3
0003 ||------] [-------------[TON 10 s]-------------------------------( )------||
          ||
          ||  Rmotor2         X1s2                                      Rt4
0004 ||------] [-------------] [----------------------------------------( )------||
          ||
          ||  Rmotor1         Xem_stop                                  Rt5
0005 ||------] [-------------] [----------------------------------------( )------||
          ||
          ||  Rp1             Xem_stop                                  Rt6
0006 ||------] [-------------] [----------------------------------------( )------||
          ||
          ||  Rmotor2         Xem_stop                                  Rt7
0007 ||------] [-------------] [----------------------------------------( )------||
```

The Places program is as follows.

```
          ||    Rmotor1             Rp1            Rmotor2            Rt1             RIdle      ||
  0008|+------]/[-----------]/[------------]/[------+------]/[--------------( )------||
          |     Rt4                                        |                                    |
          |+------] [-----------------------------------+                                    |
          |     Rt5                                        |                                    |
          |+------] [-----------------------------------+                                    |
          |     Rt6                                        |                                    |
          |+------] [-----------------------------------+                                    |
          |     Rt7                                        |                                    |
          |+------] [-----------------------------------+                                    |
          |     RIdle                                      |                                    |
          |+------] [-----------------------------------+                                    |
          ||    Rt1                Rt2            Rt5                             Rmotor1    ||
  0009|+------] [------+------]/[------------]/[-------------------------------( )------||
          |     Rmotor1    |                                                                   |
          |+------] [------+                                                                   |
          ||    Rt2                Rt3            Rt6                             Rp1        ||
  0010|+------] [------+------]/[------------]/[-------------------------------( )------||
          |     Rp1        |                                                                   |
          |+------] [------+                                                                   |
          ||    Rt3                Rt4            Rt7                             Rmotor2    ||
  0011|+------] [------+------]/[------------]/[-------------------------------( )------||
          |     Rmotor2    |                                                                   |
          |+------] [------+                                                                   |
          ||    Rt4                                                              C1         ||
  0012|+------] [------+----------------------------------------------------{RES}-----||
          |     Xem_stop   |                                                   T1         ||
          |+------] [------+----------------------------------------------------{RES}-----||
          ||                                                                                  ||
```

The Outputs program is as follows.

```
          ||    Rmotor1                                                          Ymotor     ||
  0013|+------] [------+----------------------------------------------------------( )------||
          |     Rmotor2    |                                                                  |
          |+------] [------+                                                                  |
          ||                                                                                  ||
```

To produce Arduino sketch we select settings → Microcontroller → Atmel AVR Atmega328 28-PDIP.

Compile → Compile Sketch for ARDUINO

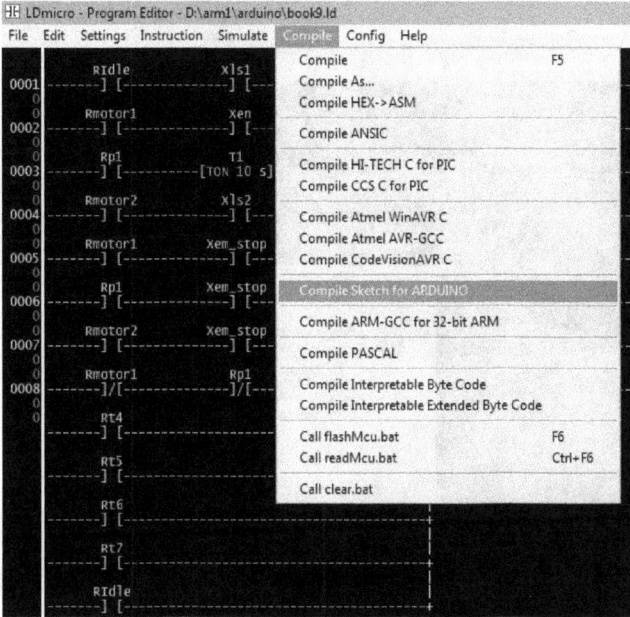

Then we save the produced Arduino sketch file as shown below.

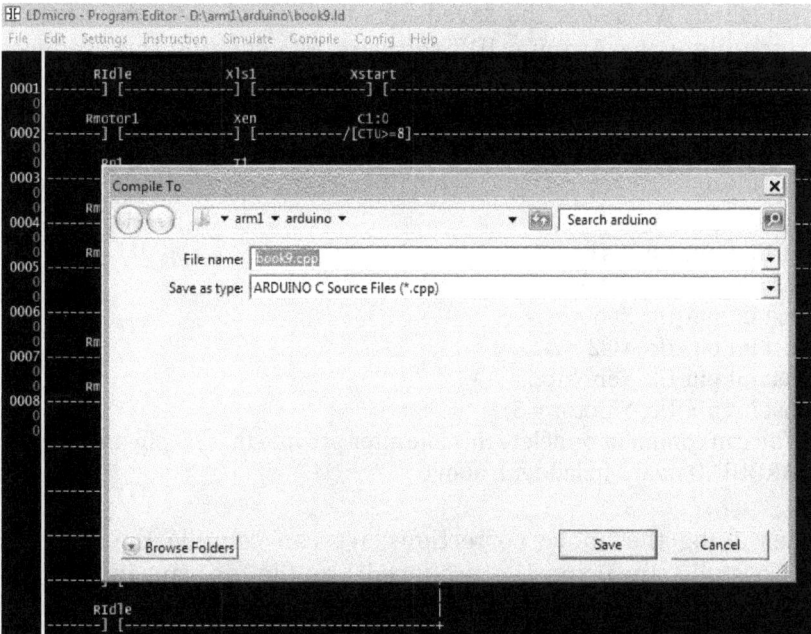

After do saving, the compile successful message is depicted as follows.

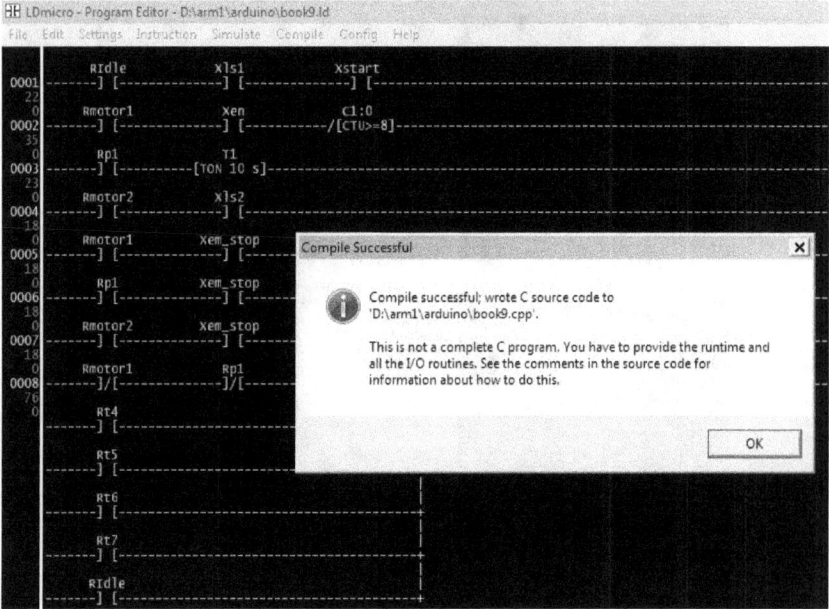

By selecting OK button the process of producing Arduino sketch file is finished. We select the saved files folder and open the Arduino sketch file using Arduino IDE software. We compile the produced Arduino sketch file in Arduino IDE environment. By compiling the program a compiling error message is displayed in ladder.h file. We must provide the I/O pin mapping for Arduino board in ladder.h file as follows.

```
const int pin_Ub_Xls1 = A1;
const int pin_Ub_Xstart = A0;
const int pin_Ub_Xen = A3;
const int pin_Ub_Xls2 = A2;
const int pin_Ub_Xem_stop = A4;
const int pin_Ub_Ymotor = 3;
//You can comment or delete this line after provide the I/O pin mapping for
//ARDUINO board in ladder.h above.
```

After doing the above corrections, we can compile the program successfully. By using the produced hex file we can apply it for simulating on Arduino Nano board using Proteus software as shown below.

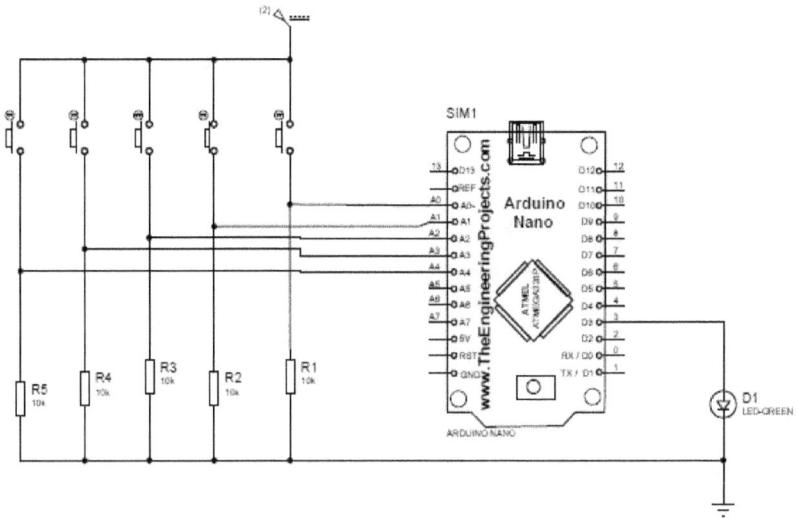

Chapter 10

Low Cost Arduino based HMI Driving Touch Screen TFT LCDs and Modbus RTU Networking

10.1 Introduction

Since, using a PC as an HMI and also HMI modules produced by famous PLC manufacturers are very expensive therefore, in this chapter we will evaluate how to build and design HMIs for Arduino based PLCs using available low cost touch screen TFT LCDs produced for Arduino boards. MCUFRIEND Company produces very cheap touch screen TFT LCDs for Arduino boards which can be used as low cost HMIs in Arduino based PLCs. An Arduino library entitled MCUFRIEND_kbv is written by David Prentice for Arduino 2.4, 2.8, 3.5, 3.6, and 3.95 inch MCUFRIEND Shields which is used extensively for MCUFRIEND Shields. Also, a library entitled GUIslice is presented by Calvin Hass for programming touch screen TFT LCDs using Arduino IDE. It has also an associated graphical design software (GUIslice builder) which can be used to produce arduino sketch file from designed HMI in GUIslice builder software automatically.

In a traditional industrial control system, the controller and the equipment is connected directly to each other, but in a PLC-based system the wiring between the controller and equipment is omitted. Because all equipment are routed through the PLC with a communication protocol and the control program in the PLC handles communicating between colligated equipment. Because Arduino is capable of dealing with digital and analog signals, it is a reasonable solution for making alternatives like analog input modules for PLCs. The old RS-232 serial communications standard is used in the industrial sector. However, to wire an Arduino (or any other micro) to a PLC, logic-level translation (TTL to RS-232) is essential because of the existing voltage level differences between TTL and RS-232 protocol. The main element is a serial communication protocol called Modbus. Modbus is an adaptive serial communication protocol developed by Modicon for use within PLCs. Modbus is transmitted over a single serial cable connecting the serial ports on two devices: a master and a slave. Modbus is a free and open communication protocol, is used easily to connect Arduino boards to PLCs. While RS-232 is the most common serial interface, RS-485 allows multiple devices (up to 32) to communicate at half-duplex on a single pair of wires (and a ground wire) up to 1,200 meters. The length of the network and the number of nodes can easily be extended using available repeaters.

Modbus RS-485 is the most flexible serial communication standard used in a wide range of computer/automation systems and within PLCs.

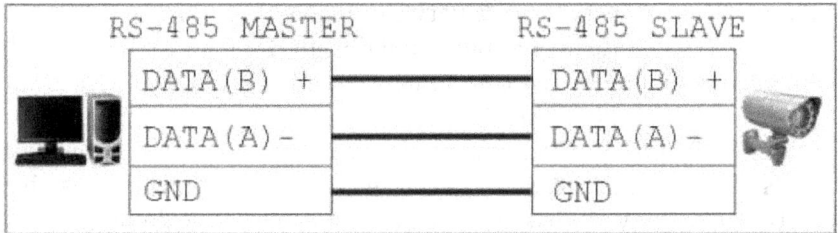

Modbus is a fairly simple protocol for data collecting from different devices for monitoring the operations and troubleshooting them from a central remote place. A protocol defines the data structure however, an electrical standard determines how the data is physically transmitted. There are many different protocols such as Modbus that can be used on RS-232/RS-485 wired systems.

10.2 Arduino based HMI using GUIslice builder

In this section we will use GUIslice builder software to make an Arduino based HMI. The GUIslice software environment is as follows. To begin we use some toolbox elements (Text, Ring Gauge, Radial Gauge and Ramp Gauge) to display an analog input value connected to the ADC pin of Arduino board which is shown below.

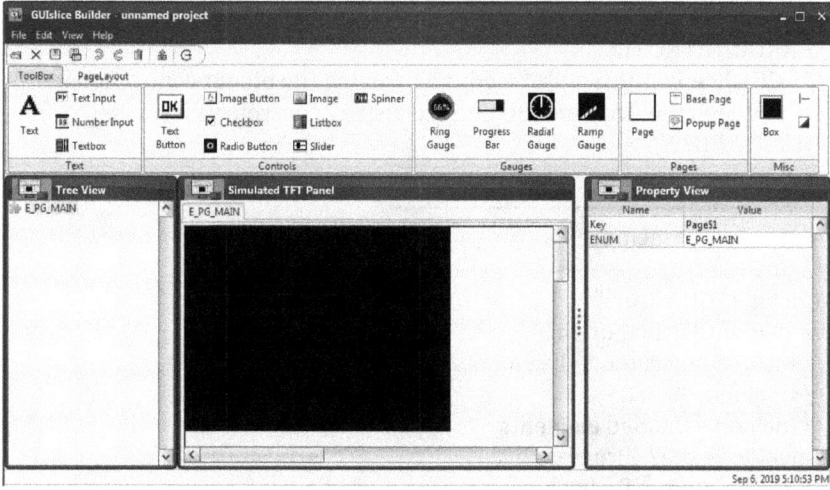

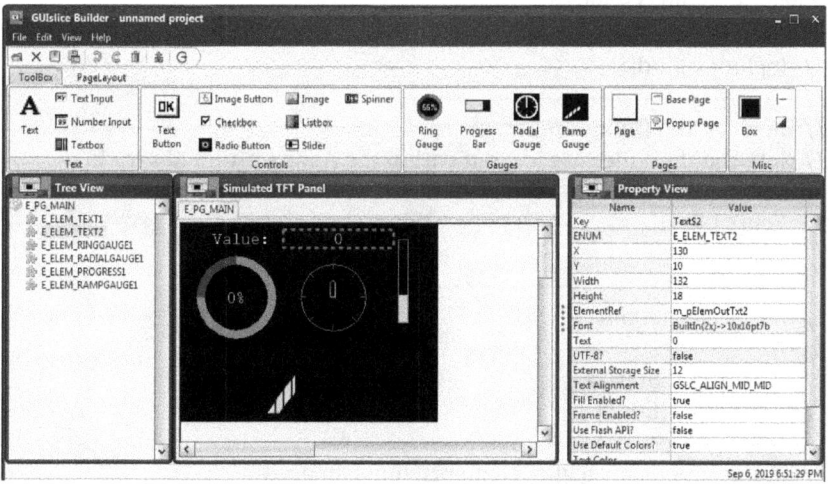

We save the project then select from file menu → Generate Code.

The generated Arduino sketch code is as follows.

```
//<File !Start!>
// FILE: [value.ino]
// Created by GUIslice Builder version: [0.13.0]
// GUIslice Builder Generated File
// For the latest guides, updates and support view:
// https://github.com/ImpulseAdventure/GUIslice
//
//<File !End!>
```

```
//
// ARDUINO NOTES:
// - GUIslice_config.h must be edited to match the pinout connections
//   between the Arduino CPU and the display controller (see
ADAGFX_PIN_*).
//
// ------------------------------------------------
// Headers to include
// ------------------------------------------------
#include "GUIslice.h"
#include "GUIslice_drv.h"
// Include any extended elements
//<Includes !Start!>
// Include extended elements
#include "elem/XProgress.h"
#include "elem/XRadial.h"
#include "elem/XRamp.h"
#include "elem/XRingGauge.h"
//<Includes !End!>
// ------------------------------------------------
// Headers and Defines for fonts
// Note that font files are located within the Adafruit-GFX library folder:
// ------------------------------------------------
//<Fonts !Start!>
//<Fonts !End!>
// ------------------------------------------------
// Defines for resources
// ------------------------------------------------
//<Resources !Start!>
//<Resources !End!>
// ------------------------------------------------
// Enumerations for pages, elements, fonts, images
// ------------------------------------------------
//<Enum !Start!>
enum {E_PG_MAIN};
enum {E_ELEM_PROGRESS1,E_ELEM_RADIALGAUGE1,E_ELEM_RAMP-
GAUGE1
    ,E_ELEM_RINGGAUGE1,E_ELEM_TEXT1,E_ELEM_TEXT2};
// Must use separate enum for fonts with MAX_FONT at end to use
gslc_FontSet.
enum {E_FONT_TXT10,MAX_FONT};
//<Enum !End!>
// ------------------------------------------------
// Instantiate the GUI
// ------------------------------------------------
```

```
// ----------------------------------------------
// Define the maximum number of elements and pages
// ----------------------------------------------
//<ElementDefines !Start!>
#define MAX_PAGE            1
#define MAX_ELEM_PG_MAIN 6                          // # Elems total on
page
#define MAX_ELEM_PG_MAIN_RAM MAX_ELEM_PG_MAIN // # Elems in
RAM
//<ElementDefines !End!>
// ----------------------------------------------
// Create element storage
// ----------------------------------------------
gslc_tsGui              m_gui;
gslc_tsDriver           m_drv;
gslc_tsFont             m_asFont[MAX_FONT];
gslc_tsPage             m_asPage[MAX_PAGE];
//<GUI_Extra_Elements !Start!>
gslc_tsElem             m_asPage1Elem[MAX_ELEM_PG_MAIN_RAM];
gslc_tsElemRef          m_asPage1ElemRef[MAX_ELEM_PG_MAIN];
gslc_tsXRingGauge       m_sXRingGauge1;
gslc_tsXRadial          m_sXRadialGauge1;
gslc_tsXProgress        m_sXBarGauge1;
gslc_tsXRamp            m_sXRampGauge1;
#define MAX_STR          100
//<GUI_Extra_Elements !End!>
// ----------------------------------------------
// Program Globals
// ----------------------------------------------
// Save some element references for direct access
//<Save_References !Start!>
gslc_tsElemRef* m_pElemOutTxt2       = NULL;
gslc_tsElemRef* m_pElemProgress1     = NULL;
gslc_tsElemRef* m_pElemRadial1       = NULL;
gslc_tsElemRef* m_pElemRamp1         = NULL;
gslc_tsElemRef* m_pElemXRingGauge1   = NULL;
//<Save_References !End!>
// Define debug message function
static int16_t DebugOut(char ch) { if (ch == (char)'\n') Serial.println("");
else Serial.write(ch); return 0; }
// ----------------------------------------------
// Callback Methods
// ----------------------------------------------
//<Button Callback !Start!>
//<Button Callback !End!>
```

```
//<Checkbox Callback !Start!>
//<Checkbox Callback !End!>
//<Keypad Callback !Start!>
//<Keypad Callback !End!>
//<Spinner Callback !Start!>
//<Spinner Callback !End!>
//<Listbox Callback !Start!>
//<Listbox Callback !End!>
//<Draw Callback !Start!>
//<Draw Callback !End!>
//<Slider Callback !Start!>
//<Slider Callback !End!>
//<Tick Callback !Start!>
//<Tick Callback !End!>
// -----------------------------------------------
// Create page elements
// -----------------------------------------------
bool InitGUI()
{
  gslc_tsElemRef* pElemRef = NULL;
//<InitGUI !Start!>
gslc_PageAdd(&m_gui,E_PG_MAIN,m_asPage1Elem,MAX_ELEM_PG_MAIN_
RAM,m_asPage1ElemRef,MAX_ELEM_PG_MAIN);

  // NOTE: The current page defaults to the first page added. Here we ex-
plicitly
  //     ensure that the main page is the correct page no matter the add or-
der.
  gslc_SetPageCur(&m_gui,E_PG_MAIN);
  // Set Background to a flat color
  gslc_SetBkgndColor(&m_gui,GSLC_COL_BLACK);
  // ----------------------------------
  // PAGE: E_PG_MAIN
  // Create E_ELEM_TEXT1 text label
  pElemRef = gslc_ElemCre-
ateTxt(&m_gui,E_ELEM_TEXT1,E_PG_MAIN,(gslc_tsRect){40,10,84,18},
    (char*)"Value: ",0,E_FONT_TXT10);
  // Create E_ELEM_TEXT2 runtime modifiable text
  static char m_sDisplayText2[13] = "0";
  pElemRef = gslc_ElemCre-
ateTxt(&m_gui,E_ELEM_TEXT2,E_PG_MAIN,(gslc_tsRect){130,10,132,18},
    (char*)m_sDisplayText2,13,E_FONT_TXT10);
  gslc_ElemSetTxtAlign(&m_gui,pElemRef,GSLC_ALIGN_MID_MID);
  m_pElemOutTxt2 = pElemRef;
  // Create ring gauge E_ELEM_RINGGAUGE1
```

```
  static char m_sRingText1[11] = "";
  pElemRef = gslc_ElemXRingGaugeCreate(&m_gui,E_ELEM_RING-
GAUGE1,E_PG_MAIN,&m_sXRingGauge1,
      (gslc_tsRect){20,40,100,100},
      (char*)m_sRingText1,11,E_FONT_TXT10);
  gslc_ElemXRingGaugeSetValRange(&m_gui, pElemRef, 0, 100);
  gslc_ElemXRingGaugeSetVal(&m_gui, pElemRef, 0); // Set initial value
  m_pElemXRingGauge1 = pElemRef;
  // Create progress bar E_ELEM_RADIALGAUGE1
  pElemRef = gslc_ElemXRadialCreate(&m_gui,E_ELEM_RADIAL-
GAUGE1,E_PG_MAIN,&m_sXRadialGauge1,
    (gslc_tsRect){150,50,80,80},0,100,0,GSLC_COL_GREEN);
  gslc_ElemXRadialSetIndica-
tor(&m_gui,pElemRef,GSLC_COL_GREEN,20,3,false);
  gslc_ElemXRadialSetTicks(&m_gui,pElemRef,GSLC_COL_GRAY,8,5);
  m_pElemRadial1 = pElemRef;
  // Create progress bar E_ELEM_PROGRESS1
  pElemRef = gslc_ElemXProgressCreate(&m_gui,E_ELEM_PRO-
GRESS1,E_PG_MAIN,&m_sXBarGauge1,
    (gslc_tsRect){270,20,12,100},0,100,0,GSLC_COL_GREEN,true);
  m_pElemProgress1 = pElemRef;
  // Create progress bar E_ELEM_RAMPGAUGE1
  pElemRef = gslc_ElemXRampCreate(&m_gui,E_ELEM_RAMP-
GAUGE1,E_PG_MAIN,&m_sXRampGauge1,
    (gslc_tsRect){110,150,100,80},0,100,
    0,GSLC_COL_YELLOW,false);
  m_pElemRamp1 = pElemRef;
//<InitGUI !End!>
  return true;
}
void setup()
{
  // -------------------------------------------------
  // Initialize
  // -------------------------------------------------
  Serial.begin(9600);
  //'Wait for USB Serial
  //delay(1000);  // NOTE: Some devices require a delay after Se-
rial.begin() before serial port can be used
  gslc_InitDebug(&DebugOut);
  if
(!gslc_Init(&m_gui,&m_drv,m_asPage,MAX_PAGE,m_asFont,MAX_FONT)) {
return; }
  // -------------------------------------------------
  // Load Fonts
```

```
// ------------------------------------------------
//<Load_Fonts !Start!>
  if (!gslc_FontSet(&m_gui,E_FONT_TXT10,GSLC_FONTREF_PTR,NULL,2))
{ return; }
//<Load_Fonts !End!>
  // ------------------------------------------------
  // Create graphic elements
  // ------------------------------------------------
  InitGUI();
//<Startup !Start!>
//<Startup !End!>
}
// ---------------------------------
// Main event loop
// ---------------------------------
void loop()
{
  char acStr[10];
  // Read the ADC
  uint16_t nVal = analogRead(A0)/10;
  sprintf(acStr,"%u",nVal);
  gslc_ElemSetTxtStr(&m_gui,m_pElemOutTxt2,acStr);
  gslc_ElemXRadialSetVal(&m_gui, m_pElemRadial1, nVal);
  gslc_ElemXProgressSetVal(&m_gui, m_pElemProgress1, nVal);
  gslc_ElemXRampSetVal(&m_gui, m_pElemRamp1, nVal);
  gslc_ElemXRingGaugeSetVal(&m_gui, m_pElemXRingGauge1, nVal);
  gslc_ElemSetTxtStr(&m_gui,m_pElemXRingGauge1,acStr);
  delay(50);
  // ------------------------------------------------
  // Update GUI Elements
  // ------------------------------------------------
  //TODO - Add update code for any text, gauges, or sliders
  // ------------------------------------------------
  // Periodically call GUIslice update function
  // ------------------------------------------------
  gslc_Update(&m_gui);
}
```

To display the analog value connected to pin A0 of Arduino we should add
the following codes in the void loop() section as shown in the above Arduino sketch code.

```
  char acStr[10];
  // Read the ADC
  uint16_t nVal = analogRead(A0)/10;
```

```
sprintf(acStr,"%u",nVal);
gslc_ElemSetTxtStr(&m_gui,m_pElemOutTxt2,acStr);
gslc_ElemXRadialSetVal(&m_gui, m_pElemRadial1, nVal);
gslc_ElemXProgressSetVal(&m_gui, m_pElemProgress1, nVal);
gslc_ElemXRampSetVal(&m_gui, m_pElemRamp1, nVal);
gslc_ElemXRingGaugeSetVal(&m_gui, m_pElemXRingGauge1, nVal);
gslc_ElemSetTxtStr(&m_gui,m_pElemXRingGauge1,acStr);
delay(50);
```

Using the generated hex code, we can simulate the designed HMI with Arduino Mega 2560 and ILI9341 TFT shield in Proteus software as illustrated below.

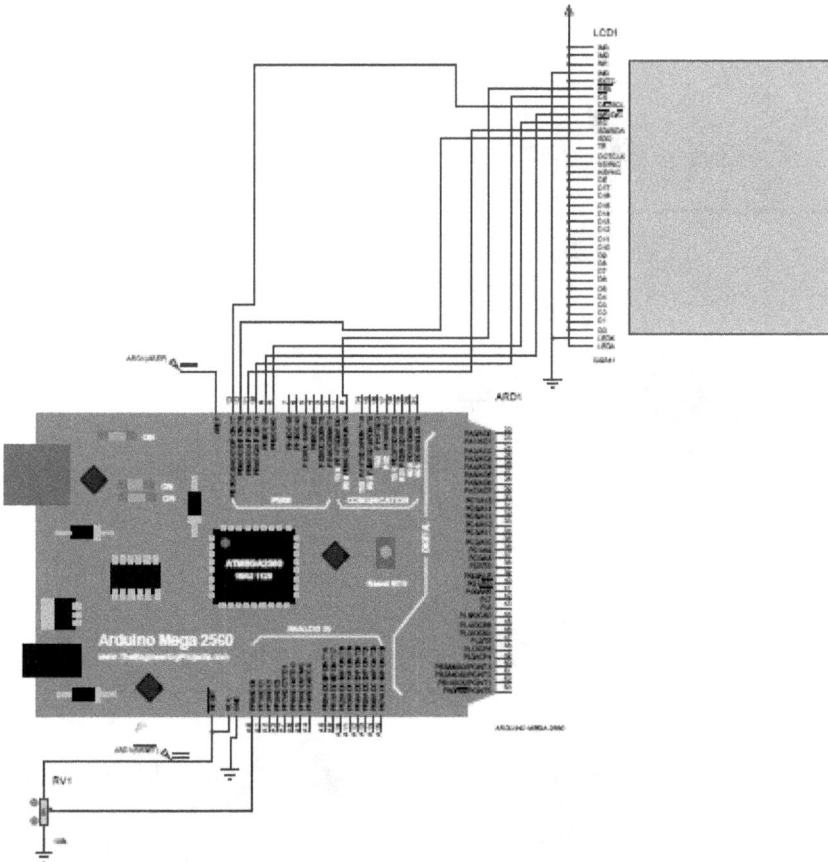

10.3 Arduino Modbus RTU networking

Arduino based Modbus RTU networking is discussed in this section. A Modbus library presented by Jose Maria is used for Modbus RTU networking in Arduino boards. This Arduino IDE library could be found in the following website.
https://github.com/pepsilla/Arduino

To explain Modbus RTU networking in Arduino boards we consider the following system implemented in Proteus software.

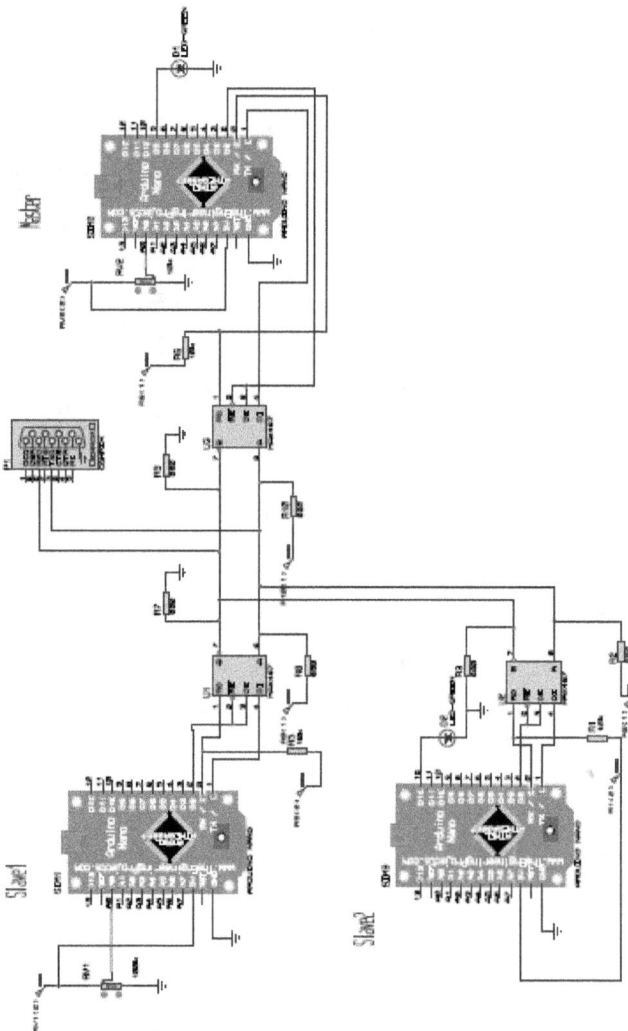

The Arduino sketch programs for master and slaves are as follows. The master program is given below.

```
#include <SimpleModbusMaster.h>
/*
  The example will use packet1 to read a register from address 0 (the adc
  ch0 value)
  from the arduino slave (id=1). It will then use this value to adjust the
  brightness
  of an led on pin 9 using PWM.
  It will then use packet2 to write a register (its own adc ch0 value) to ad-
  dress 1
  on the arduino slave (id=1) adjusting the brightness of an led on pin 9 us-
  ing PWM.
*/
////////////////////// Port information //////////////////////
#define baud 9600
#define timeout 1000
#define polling 200 // the scan rate
#define retry_count 10
// used to toggle the receive/transmit pin on the driver
#define TxEnablePin 2
#define LED 9
// The total amount of available memory on the master to store data
#define TOTAL_NO_OF_REGISTERS 1
// This is the easiest way to create new packets
// Add as many as you want. TOTAL_NO_OF_PACKETS
// is automatically updated.
enum
{
  PACKET1,
  PACKET2,
  TOTAL_NO_OF_PACKETS // leave this last entry
};
// Create an array of Packets to be configured
Packet packets[TOTAL_NO_OF_PACKETS];
// Masters register array
unsigned int regs[TOTAL_NO_OF_REGISTERS];
void setup()
{
  // Initialize each packet
  //---------------pcket arry pointer, node ID, Function,  adress,data,lo-
cal_adress_start
  modbus_construct(&packets[PACKET1], 1, READ_HOLDING_REGISTERS,
0, 1, 0);
```

```
  modbus_construct(&packets[PACKET2], 2, PRESET_MULTIPLE_REGIS-
TERS, 1, 1, 0);
  // Initialize the Modbus Finite State Machine
  modbus_configure(&Serial, baud, SERIAL_8N1, timeout, polling, re-
try_count, TxEnablePin, packets, TOTAL_NO_OF_PACKETS, regs);
  pinMode(LED, OUTPUT);
}
void loop()
{
  modbus_update();
  // regs[0] = analogRead(0); // update data to be written to arduino slave
  //if (reg[0] > 200) {
  //  digitalWrite (LED, HIGH);
  //} else {
  //  digitalWrite (LED, LOW);
  //}
  analogWrite(LED, regs[0]>>2); // constrain adc value from the arduino
slave to 255
  regs[0] = analogRead(0);
}
```

The slave1 program is as follows.

```
#include <SimpleModbusSlave.h>
/*
  SimpleModbusSlaveV10 supports function 3, 6 & 16.
    This example code will receive the adc ch0 value from the arduino mas-
ter.
  It will then use this value to adjust the brightness of the led on pin 9.
  The value received from the master will be stored in address 1 in its own
  address space namely holdingRegs[].

  In addition to this the slaves own adc ch0 value will be stored in
  address 0 in its own address space holdingRegs[] for the master to
  be read. The master will use this value to alter the brightness of its
  own led connected to pin 9.
  The modbus_update() method updates the holdingRegs register array and
checks
  communication.
  Note:
  The Arduino serial ring buffer is 64 bytes or 32 registers.
  Most of the time you will connect the arduino to a master via serial
  using a MAX485 or similar.
  In a function 3 request the master will attempt to read from your
  slave and since 5 bytes is already used for ID, FUNCTION, NO OF BYTES
```

and two BYTES CRC the master can only request 58 bytes or 29 registers.
In a function 16 request the master will attempt to write to your
slave and since a 9 bytes is already used for ID, FUNCTION, ADDRESS,
NO OF REGISTERS, NO OF BYTES and two BYTES CRC the master can only
write
54 bytes or 27 registers.
Using a USB to Serial converter the maximum bytes you can send is
limited to its internal buffer which differs between manufactures.
*/

```
// Using the enum instruction allows for an easy method for adding and
// removing registers. Doing it this way saves you #defining the size
// of your slaves register array each time you want to add more registers
// and at a glimpse informs you of your slaves register layout.
#define  ledPin 12
//#define  button 11
/////////////// registers of your slave /////////////////
enum
{
  // just add or remove registers and your good to go...
  // The first register starts at address 0
  ADC_VAL,
  // OUT_VAL,
  // LED_VAL,
  // IND_VAL,
  HOLDING_REGS_SIZE // leave this one
  // total number of registers for function 3 and 16 share the same register
array
  // i.e. the same address space
};
unsigned int holdingRegs[HOLDING_REGS_SIZE]; // function 3 and 16 register array
/////////////////////////////////////////////////////////
void setup()
{
  /* parameters(HardwareSerial* SerialPort,
         long baudrate,
                 unsigned char byteFormat,
         unsigned char ID,
         unsigned char transmit enable pin,
         unsigned int holding registers size,
         unsigned int* holding register array)
  */
  /* Valid modbus byte formats are:
  SERIAL_8N2: 1 start bit, 8 data bits, 2 stop bits
  SERIAL_8E1: 1 start bit, 8 data bits, 1 Even parity bit, 1 stop bit
```

SERIAL_8O1: 1 start bit, 8 data bits, 1 Odd parity bit, 1 stop bit
You can obviously use SERIAL_8N1 but this does not adhere to the
Modbus specifications. That said, I have tested the SERIAL_8N1 option
on various commercial masters and slaves that were suppose to adhere
to this specification and was always able to communicate... Go figure.
These byte formats are already defined in the Arduino global name space.

```
*/
modbus_configure(&Serial, 9600, SERIAL_8N1, 1, 2, HOLDING_REGS_SIZE,
holdingRegs);
// modbus_update_comms(baud, byteFormat, id) is not needed but allows
for easy update of the
// port variables and slave id dynamically in any function.
modbus_update_comms(9600, SERIAL_8N1, 1);
pinMode(ledPin, OUTPUT);
}
void loop()
{
// modbus_update() is the only method used in loop(). It returns the total
error
// count since the slave started. You don't have to use it but it's useful
// for fault finding by the modbus master.
modbus_update();
holdingRegs[ADC_VAL] = analogRead(A0);
//holdingRegs[OUT_VAL] = analogRead(A1);
//holdingRegs[IND_VAL] = digitalRead(button);
//if (holdingRegs[LED_VAL] == 1) {
// turn LED on:
// digitalWrite(ledPin, HIGH);
//} else {
// turn LED off:
// digitalWrite(ledPin, LOW);
//}
/* Note:
The use of the enum instruction is not needed. You could set a maximum
allowable
size for holdinRegs[] by defining HOLDING_REGS_SIZE using a constant
and then access
holdingRegs[] by "Index" addressing.
I.e.
holdingRegs[0] = analogRead(A0);
analogWrite(LED, holdingRegs[1]/4);
*/
}
```

The slave2 program is given as below.

```
#include <SimpleModbusSlave.h>
/*
   SimpleModbusSlaveV10 supports function 3, 6 & 16.
   This example code will receive the adc ch0 value from the arduino master.
   It will then use this value to adjust the brightness of the led on pin 9.
   The value received from the master will be stored in address 1 in its own
   address space namely holdingRegs[].
   In addition to this the slaves own adc ch0 value will be stored in
   address 0 in its own address space holdingRegs[] for the master to
   be read. The master will use this value to alter the brightness of its
   own led connected to pin 9.
   The modbus_update() method updates the holdingRegs register array and
checks
   communication.
   Note:
   The Arduino serial ring buffer is 64 bytes or 32 registers.
   Most of the time you will connect the arduino to a master via serial
   using a MAX485 or similar.
   In a function 3 request the master will attempt to read from your
   slave and since 5 bytes is already used for ID, FUNCTION, NO OF BYTES
   and two BYTES CRC the master can only request 58 bytes or 29 registers.
   In a function 16 request the master will attempt to write to your
   slave and since a 9 bytes is already used for ID, FUNCTION, ADDRESS,
   NO OF REGISTERS, NO OF BYTES and two BYTES CRC the master can only
write
   54 bytes or 27 registers.
   Using a USB to Serial converter the maximum bytes you can send is
   limited to its internal buffer which differs between manufactures.
*/
// Using the enum instruction allows for an easy method for adding and
// removing registers. Doing it this way saves you #defining the size
// of your slaves register array each time you want to add more registers
// and at a glimpse informs you of your slaves register layout.
#define  ledPin 12
//#define  button 11
/////////////////// registers of your slave ///////////////////
enum
{
  // just add or remove registers and your good to go...
  // The first register starts at address 0
  PWM_VAL,
  ADC2_VAL,
  // OUT_VAL,
```

```
// LED_VAL,
// IND_VAL,
 HOLDING_REGS_SIZE // leave this one
 // total number of registers for function 3 and 16 share the same register
array
 // i.e. the same address space
};
unsigned int holdingRegs[HOLDING_REGS_SIZE]; // function 3 and 16 reg-
ister array
///////////////////////////////////////////////////////////
void setup()
{
  /* parameters(HardwareSerial* SerialPort,
        long baudrate,
                unsigned char byteFormat,
        unsigned char ID,
        unsigned char transmit enable pin,
        unsigned int holding registers size,
        unsigned int* holding register array)
  */
  /* Valid modbus byte formats are:
    SERIAL_8N2: 1 start bit, 8 data bits, 2 stop bits
    SERIAL_8E1: 1 start bit, 8 data bits, 1 Even parity bit, 1 stop bit
    SERIAL_8O1: 1 start bit, 8 data bits, 1 Odd parity bit, 1 stop bit
    You can obviously use SERIAL_8N1 but this does not adhere to the
    Modbus specifications. That said, I have tested the SERIAL_8N1 option
    on various commercial masters and slaves that were suppose to adhere
    to this specification and was always able to communicate... Go figure.
    These byte formats are already defined in the Arduino global name space.
  */
  modbus_configure(&Serial, 9600, SERIAL_8N1, 2, 2, HOLDING_REGS_SIZE,
holdingRegs);

  // modbus_update_comms(baud, byteFormat, id) is not needed but allows
for easy update of the
  // port variables and slave id dynamically in any function.
  modbus_update_comms(9600, SERIAL_8N1, 2);
  pinMode(ledPin, OUTPUT);
}
void loop()
{
  // modbus_update() is the only method used in loop(). It returns the total
error
  // count since the slave started. You don't have to use it but it's useful
  // for fault finding by the modbus master.
```

```
modbus_update();
//holdingRegs[ADC_VAL] = analogRead(A0);
//holdingRegs[OUT_VAL] = analogRead(A1);
//holdingRegs[IND_VAL] = digitalRead(button);
//if (holdingRegs[LED_VAL] == 1) {
 // turn LED on:
 // digitalWrite(ledPin, HIGH);
//} else {
 // turn LED off:
 // digitalWrite(ledPin, LOW);
// }
if (holdingRegs[ADC2_VAL] > 100) {
 // turn LED on:
 digitalWrite(ledPin, HIGH);
} else {
 // turn LED off:
 digitalWrite(ledPin, LOW);
}
/* Note:
 The use of the enum instruction is not needed. You could set a maximum allowable
 size for holdinRegs[] by defining HOLDING_REGS_SIZE using a constant and then access
 holdingRegs[] by "Index" addressing.
 I.e.
 holdingRegs[0] = analogRead(A0);
 analogWrite(LED, holdingRegs[1]/4);
*/
}
```

10.4 SCADA LAquis Modbus RTU software

Arduino boards as Modbus RTU slaves could communicate by Modbus RTU protocol with SCADA LAquis Modbus RTU software as a master. For example we consider the following system designed with Proteus software.

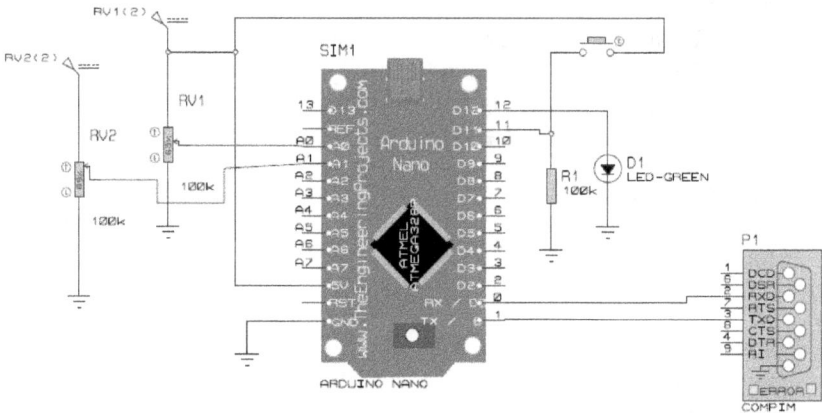

The following Arduino sketch Modbus RTU slave program is used to communicate with SCADA LAquis Modbus RTU protocol.

```
#include <SimpleModbusSlave.h>

/*
SimpleModbusSlaveV10 supports function 3, 6 & 16.

This example code will receive the adc ch0 value from the arduino master.
It will then use this value to adjust the brightness of the led on pin 9.
The value received from the master will be stored in address 1 in its own
address space namely holdingRegs[].

In addition to this the slaves own adc ch0 value will be stored in
address 0 in its own address space holdingRegs[] for the master to
be read. The master will use this value to alter the brightness of its
own led connected to pin 9.
The modbus_update() method updates the holdingRegs register array and
checks
communication.
Note:
The Arduino serial ring buffer is 64 bytes or 32 registers.
Most of the time you will connect the arduino to a master via serial
using a MAX485 or similar.
In a function 3 request the master will attempt to read from your
slave and since 5 bytes is already used for ID, FUNCTION, NO OF BYTES
and two BYTES CRC the master can only request 58 bytes or 29 registers.
In a function 16 request the master will attempt to write to your
slave and since a 9 bytes is already used for ID, FUNCTION, ADDRESS,
```

```
NO OF REGISTERS, NO OF BYTES and two BYTES CRC the master can only write
54 bytes or 27 registers.
Using a USB to Serial converter the maximum bytes you can send is
limited to its internal buffer which differs between manufactures.
*/
// Using the enum instruction allows for an easy method for adding and
// removing registers. Doing it this way saves you #defining the size
// of your slaves register array each time you want to add more registers
// and at a glimpse informs you of your slaves register layout.
#define  ledPin 12
#define  button 11
/////////////// registers of your slave ///////////////////
enum
{
  // just add or remove registers and your good to go...
  // The first register starts at address 0
  ADC2_VAL,
  OUT2_VAL,
  LED2_VAL,
  IND2_VAL,
  HOLDING_REGS_SIZE // leave this one
  // total number of registers for function 3 and 16 share the same register
array
  // i.e. the same address space
};
unsigned int holdingRegs[HOLDING_REGS_SIZE]; // function 3 and 16 register array
/////////////////////////////////////////////////////////////
void setup()
{
  /* parameters(HardwareSerial* SerialPort,
            long baudrate,
                  unsigned char byteFormat,
            unsigned char ID,
            unsigned char transmit enable pin,
            unsigned int holding registers size,
            unsigned int* holding register array)
  */
  /* Valid modbus byte formats are:
    SERIAL_8N2: 1 start bit, 8 data bits, 2 stop bits
    SERIAL_8E1: 1 start bit, 8 data bits, 1 Even parity bit, 1 stop bit
    SERIAL_8O1: 1 start bit, 8 data bits, 1 Odd parity bit, 1 stop bit
    You can obviously use SERIAL_8N1 but this does not adhere to the
    Modbus specifications. That said, I have tested the SERIAL_8N1 option
```

on various commercial masters and slaves that were suppose to adhere
to this specification and was always able to communicate... Go figure.
These byte formats are already defined in the Arduino global name space.
```
*/
 modbus_configure(&Serial, 9600, SERIAL_8N1, 2, 2, HOLDING_REGS_SIZE,
holdingRegs);
 // modbus_update_comms(baud, byteFormat, id) is not needed but allows
for easy update of the
 // port variables and slave id dynamically in any function.
 modbus_update_comms(9600, SERIAL_8N1, 2);
 pinMode(ledPin, OUTPUT);
}
void loop()
{
 // modbus_update() is the only method used in loop(). It returns the total
error
 // count since the slave started. You don't have to use it but it's useful
 // for fault finding by the modbus master.
 modbus_update();
 holdingRegs[ADC2_VAL] = analogRead(A0);
 holdingRegs[OUT2_VAL] = analogRead(A1);
 holdingRegs[IND2_VAL] = digitalRead(button);
 if (holdingRegs[LED2_VAL] == 1) {
  // turn LED on:
  digitalWrite(ledPin, HIGH);
 } else {
  // turn LED off:
  digitalWrite(ledPin, LOW);
 }
 /* Note:
  The use of the enum instruction is not needed. You could set a maximum
allowable
  size for holdinRegs[] by defining HOLDING_REGS_SIZE using a constant
and then access
  holdingRegs[] by "Index" addressing.
  I.e.
  holdingRegs[0] = analogRead(A0);
  analogWrite(LED, holdingRegs[1]/4);
 */
}
```

In SCADA LAquis environment, we place two analog gauges a button
(LED2) and an LED (IND2). Then we define four tags i.e. ADC2, OUT2,
LED2 and IND2. We click on Driver/PLC tab then select Modbus RTU
and select Parameter 1 (addresses) 40x0000, 40x0001, 40x0002 and

40x0003 for tags respectively. Moreover, since in the Arduino sketch program we used slave ID 2 for Arduino Nano board. We apply the value 2 in parameter 2. By using a virtual serial port driver software we can simulate the operation of this Modbus RTU based HMI successfully.

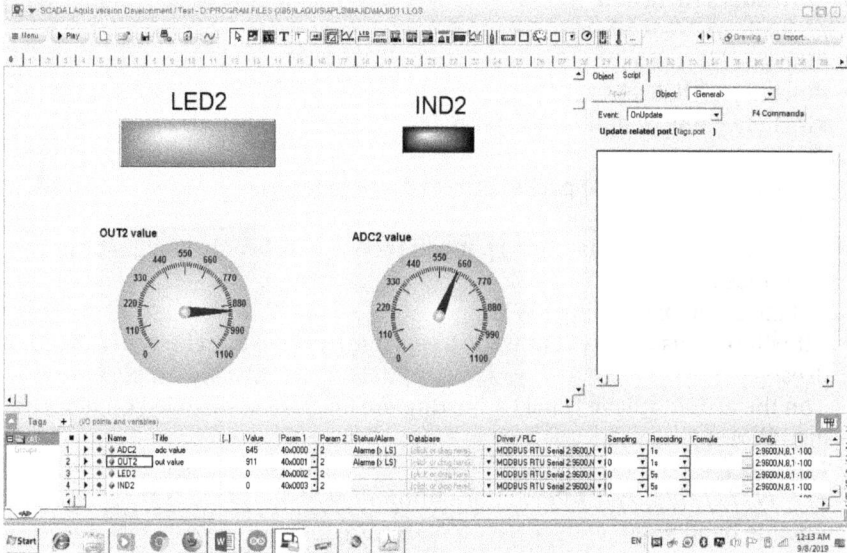

10.5 Modbus RTU HMI using low cost touch screen TFTs

In this section we combine systems used in 10-2 and 10-3 sections to implement Modbus RTU based HMI using low cost touch screen TFTs for example TFTs with driver ILI9341. Therefore, we consider the following system. Here, Arduino Mega2560 board is used as a master connected to an ILI9341 TFT. Moreover, two Arduino nano boards are used as slaves as described in section 10-3. The Arduino sketch program for Arduino Mega2560 board as a master is written as follows.

```
//<File !Start!>
// FILE: [value.ino]
// Created by GUIslice Builder version: [0.13.0]
// GUIslice Builder Generated File
// For the latest guides, updates and support view:
// https://github.com/ImpulseAdventure/GUIslice
```

```
//<File !End!>
// ARDUINO NOTES:
// - GUIslice_config.h must be edited to match the pinout connections
//   between the Arduino CPU and the display controller (see
ADAGFX_PIN_*).
// ------------------------------------------------
// Headers to include
// ------------------------------------------------
#include "GUIslice.h"
#include "GUIslice_drv.h"
#include <SimpleModbusMaster.h>
/*
   The example will use packet1 to read a register from address 0 (the adc
ch0 value)
   from the arduino slave (id=1). It will then use this value to adjust the
brightness
   of an led on pin 9 using PWM.
   It will then use packet2 to write a register (its own adc ch0 value) to ad-
dress 1
   on the arduino slave (id=1) adjusting the brightness of an led on pin 9 us-
ing PWM.
*/
///////////////////// Port information /////////////////////
#define baud 9600
#define timeout 1000
#define polling 200 // the scan rate
#define retry_count 10
// used to toggle the receive/transmit pin on the driver
#define TxEnablePin 2
//#define LED 3
// The total amount of available memory on the master to store data
#define TOTAL_NO_OF_REGISTERS 1
// This is the easiest way to create new packets
// Add as many as you want. TOTAL_NO_OF_PACKETS
// is automatically updated.
enum
{
  PACKET1,
  PACKET2,
  TOTAL_NO_OF_PACKETS // leave this last entry
};
// Create an array of Packets to be configured
Packet packets[TOTAL_NO_OF_PACKETS];
// Masters register array
unsigned int regs[TOTAL_NO_OF_REGISTERS];
```

```
// Include any extended elements
//<Includes !Start!>
// Include extended elements
#include "elem/XProgress.h"
#include "elem/XRadial.h"
#include "elem/XRamp.h"
#include "elem/XRingGauge.h"
//<Includes !End!>
// -------------------------------------------------
// Headers and Defines for fonts
// Note that font files are located within the Adafruit-GFX library folder:
// -------------------------------------------------
//<Fonts !Start!>
//<Fonts !End!>
// -------------------------------------------------
// Defines for resources
// -------------------------------------------------
//<Resources !Start!>
//<Resources !End!>
// -------------------------------------------------
// Enumerations for pages, elements, fonts, images
// -------------------------------------------------
//<Enum !Start!>
enum {E_PG_MAIN};
enum {E_ELEM_PROGRESS1,E_ELEM_RADIALGAUGE1,E_ELEM_RAMP-
GAUGE1
    ,E_ELEM_RINGGAUGE1,E_ELEM_TEXT1,E_ELEM_TEXT2};
// Must use separate enum for fonts with MAX_FONT at end to use
gslc_FontSet.
enum {E_FONT_TXT10,MAX_FONT};
//<Enum !End!>
// -------------------------------------------------
// Instantiate the GUI
// -------------------------------------------------
// -------------------------------------------------
// Define the maximum number of elements and pages
// -------------------------------------------------
//<ElementDefines !Start!>
#define MAX_PAGE          1
#define MAX_ELEM_PG_MAIN 6                    // # Elems total on
page
#define MAX_ELEM_PG_MAIN_RAM MAX_ELEM_PG_MAIN // # Elems in
RAM
//<ElementDefines !End!>
// -------------------------------------------------
```

```
// Create element storage
// -------------------------------------------------
gslc_tsGui                 m_gui;
gslc_tsDriver              m_drv;
gslc_tsFont                m_asFont[MAX_FONT];
gslc_tsPage                m_asPage[MAX_PAGE];
//<GUI_Extra_Elements !Start!>
gslc_tsElem                m_asPage1Elem[MAX_ELEM_PG_MAIN_RAM];
gslc_tsElemRef             m_asPage1ElemRef[MAX_ELEM_PG_MAIN];
gslc_tsXRingGauge          m_sXRingGauge1;
gslc_tsXRadial             m_sXRadialGauge1;
gslc_tsXProgress           m_sXBarGauge1;
gslc_tsXRamp               m_sXRampGauge1;
#define MAX_STR            100
//<GUI_Extra_Elements !End!>
// -------------------------------------------------
// Program Globals
// -------------------------------------------------
// Save some element references for direct access
//<Save_References !Start!>
gslc_tsElemRef* m_pElemOutTxt2          = NULL;
gslc_tsElemRef* m_pElemProgress1        = NULL;
gslc_tsElemRef* m_pElemRadial1          = NULL;
gslc_tsElemRef* m_pElemRamp1            = NULL;
gslc_tsElemRef* m_pElemXRingGauge1      = NULL;
//<Save_References !End!>
// Define debug message function
static int16_t DebugOut(char ch) { if (ch == (char)'\n') Serial.println("");
else Serial.write(ch); return 0; }
// -------------------------------------------------
// Callback Methods
// -------------------------------------------------
//<Button Callback !Start!>
//<Button Callback !End!>
//<Checkbox Callback !Start!>
//<Checkbox Callback !End!>
//<Keypad Callback !Start!>
//<Keypad Callback !End!>
//<Spinner Callback !Start!>
//<Spinner Callback !End!>
//<Listbox Callback !Start!>
//<Listbox Callback !End!>
//<Draw Callback !Start!>
//<Draw Callback !End!>
//<Slider Callback !Start!>
```

```
//<Slider Callback !End!>
//<Tick Callback !Start!>
//<Tick Callback !End!>
// ------------------------------------------------
// Create page elements
// ------------------------------------------------
bool InitGUI()
{
  gslc_tsElemRef* pElemRef = NULL;
//<InitGUI !Start!>
gslc_PageAdd(&m_gui,E_PG_MAIN,m_asPage1Elem,MAX_ELEM_PG_MAIN_
RAM,m_asPage1ElemRef,MAX_ELEM_PG_MAIN);
  // NOTE: The current page defaults to the first page added. Here we ex-
plicitly
  //    ensure that the main page is the correct page no matter the add or-
der.
  gslc_SetPageCur(&m_gui,E_PG_MAIN);
  // Set Background to a flat color
  gslc_SetBkgndColor(&m_gui,GSLC_COL_BLACK);
  // ----------------------------------
  // PAGE: E_PG_MAIN
  // Create E_ELEM_TEXT1 text label
  pElemRef = gslc_ElemCre-
ateTxt(&m_gui,E_ELEM_TEXT1,E_PG_MAIN,(gslc_tsRect){40,10,84,18},
    (char*)"Value: ",0,E_FONT_TXT10);
  // Create E_ELEM_TEXT2 runtime modifiable text
  static char m_sDisplayText2[13] = "0";
  pElemRef = gslc_ElemCre-
ateTxt(&m_gui,E_ELEM_TEXT2,E_PG_MAIN,(gslc_tsRect){130,10,132,18},
    (char*)m_sDisplayText2,13,E_FONT_TXT10);
  gslc_ElemSetTxtAlign(&m_gui,pElemRef,GSLC_ALIGN_MID_MID);
  m_pElemOutTxt2 = pElemRef;
  // Create ring gauge E_ELEM_RINGGAUGE1
  static char m_sRingText1[11] = "";
  pElemRef = gslc_ElemXRingGaugeCreate(&m_gui,E_ELEM_RING-
GAUGE1,E_PG_MAIN,&m_sXRingGauge1,
      (gslc_tsRect){20,40,100,100},
      (char*)m_sRingText1,11,E_FONT_TXT10);
  gslc_ElemXRingGaugeSetValRange(&m_gui, pElemRef, 0, 100);
  gslc_ElemXRingGaugeSetVal(&m_gui, pElemRef, 0); // Set initial value
  m_pElemXRingGauge1 = pElemRef;
  // Create progress bar E_ELEM_RADIALGAUGE1
  pElemRef = gslc_ElemXRadialCreate(&m_gui,E_ELEM_RADIAL-
GAUGE1,E_PG_MAIN,&m_sXRadialGauge1,
    (gslc_tsRect){150,50,80,80},0,100,0,GSLC_COL_GREEN);
```

```
 gslc_ElemXRadialSetIndica-
tor(&m_gui,pElemRef,GSLC_COL_GREEN,20,3,false);
 gslc_ElemXRadialSetTicks(&m_gui,pElemRef,GSLC_COL_GRAY,8,5);
 m_pElemRadial1 = pElemRef;
 // Create progress bar E_ELEM_PROGRESS1
 pElemRef = gslc_ElemXProgressCreate(&m_gui,E_ELEM_PRO-
GRESS1,E_PG_MAIN,&m_sXBarGauge1,
   (gslc_tsRect){270,20,12,100},0,100,0,GSLC_COL_GREEN,true);
 m_pElemProgress1 = pElemRef;
 // Create progress bar E_ELEM_RAMPGAUGE1
 pElemRef = gslc_ElemXRampCreate(&m_gui,E_ELEM_RAMP-
GAUGE1,E_PG_MAIN,&m_sXRampGauge1,
   (gslc_tsRect){110,150,100,80},0,100,
   0,GSLC_COL_YELLOW,false);
 m_pElemRamp1 = pElemRef;
//<InitGUI !End!>
 return true;
}
void setup()
{
 // -----------------------------------------------
 // Initialize
 // -----------------------------------------------
 Serial.begin(9600);
 // Initialize each packet
 //---------------pcket arry pointer, node ID, Function,  adress,data,lo-
cal_adress_start
 modbus_construct(&packets[PACKET1], 1, READ_HOLDING_REGISTERS,
0, 1, 0);
 modbus_construct(&packets[PACKET2], 2, PRESET_MULTIPLE_REGIS-
TERS, 1, 1, 0);
 // Initialize the Modbus Finite State Machine
 modbus_configure(&Serial, baud, SERIAL_8N1, timeout, polling, re-
try_count, TxEnablePin, packets, TOTAL_NO_OF_PACKETS, regs);
 // pinMode(LED, OUTPUT);
 // Wait for USB Serial
 //delay(1000); // NOTE: Some devices require a delay after Se-
rial.begin() before serial port can be used
 gslc_InitDebug(&DebugOut);
 if
(!gslc_Init(&m_gui,&m_drv,m_asPage,MAX_PAGE,m_asFont,MAX_FONT)) {
return; }
 // -----------------------------------------------
 // Load Fonts
 // -----------------------------------------------
```

```
//<Load_Fonts !Start!>
  if (!gslc_FontSet(&m_gui,E_FONT_TXT10,GSLC_FONTREF_PTR,NULL,2))
{ return; }
//<Load_Fonts !End!>
  // -----------------------------------------------
  // Create graphic elements
  // -----------------------------------------------
  InitGUI();
//<Startup !Start!>
//<Startup !End!>
}
// ---------------------------------
// Main event loop
// ---------------------------------
void loop()
{
  modbus_update();
  //analogWrite(LED, regs[0]>>2); // constrain adc value from the arduino
slave to 255
  uint16_t nVal = regs[0]/10;
  regs[0] = analogRead(2);
  char acStr[10];
  // Read the ADC
  sprintf(acStr,"%u",nVal);
  gslc_ElemSetTxtStr(&m_gui,m_pElemOutTxt2,acStr);
  gslc_ElemXRadialSetVal(&m_gui, m_pElemRadial1, nVal);
  gslc_ElemXProgressSetVal(&m_gui, m_pElemProgress1, nVal);
  gslc_ElemXRampSetVal(&m_gui, m_pElemRamp1, nVal);
  gslc_ElemXRingGaugeSetVal(&m_gui, m_pElemXRingGauge1, nVal);
  gslc_ElemSetTxtStr(&m_gui,m_pElemXRingGauge1,acStr);
  delay(500);
  // -----------------------------------------------
  // Update GUI Elements
  // -----------------------------------------------
  //TODO - Add update code for any text, gauges, or sliders

  // -----------------------------------------------
  // Periodically call GUIslice update function
  // -----------------------------------------------
  gslc_Update(&m_gui);
}
```

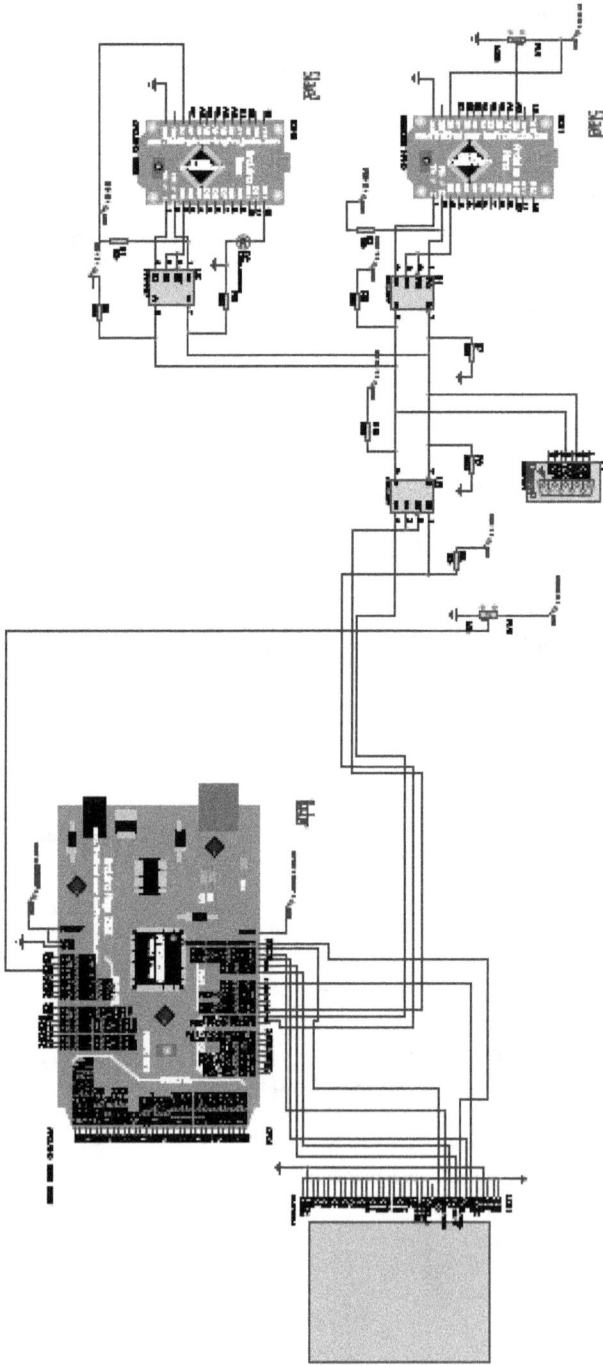

Chapter 11

Industrial Control Issues

11.1 Introduction

This chapter presents some of the real issues of industrial control problems. The reader is expected to readily solve the following problems after studying previous chapters, especially chapters 3 and 4, and the methods used (FSM diagram method and Petri nets method using PetriLLD software).

11.2 Problems

In the remainder of this chapter, 14 control problems are presented, two of which have been solved using the FSM diagram and the Petri nets method in previous chapters. The rest of the problems are left to the reader to solve them with FSM diagram/Petri nets methods.

Problem 1:

The control system works when the master switch is on. When a person enters or exits B1 or B2 sensors are activated and the door opens immediately. The door stops when the S2 switch is triggered. If none of the B1 or B2 sensors are activated, the door closes after 5 seconds and the door stops when the S1 switch is triggered. If either of the B1 or B2 sensors are activated when the door is moving, the door stops and the door opens immediately and the door stops when the S2 switch is triggered. The control system will not work when the master switch is off. Design a LAD to control this system.

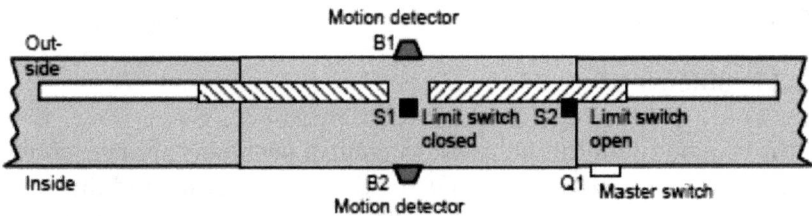

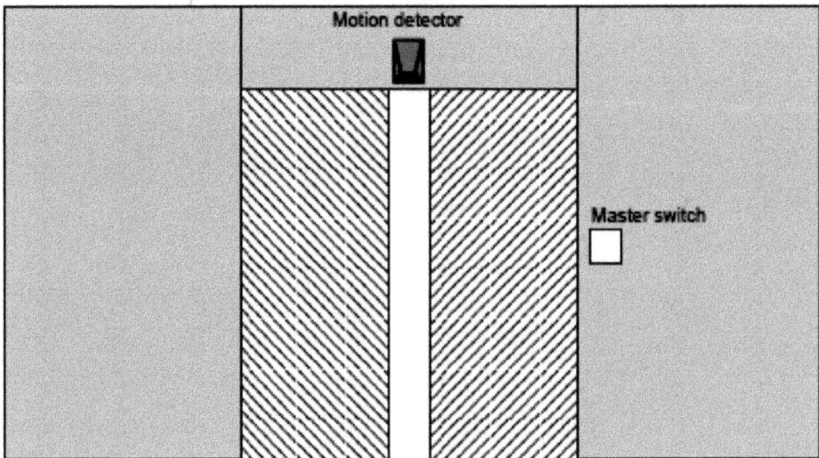

Problem 2:

(Traffic light) By pressing the start button, the G lamp illuminates for 20 seconds, then the Y lamp illuminates for 5 seconds and then the R lamp illuminates for 20 seconds, and this cycle continues. By pressing the stop button, all three lamps are switched off at any time and the start button must be pressed again to start. Design a LAD to control this system.

Traffic Lights

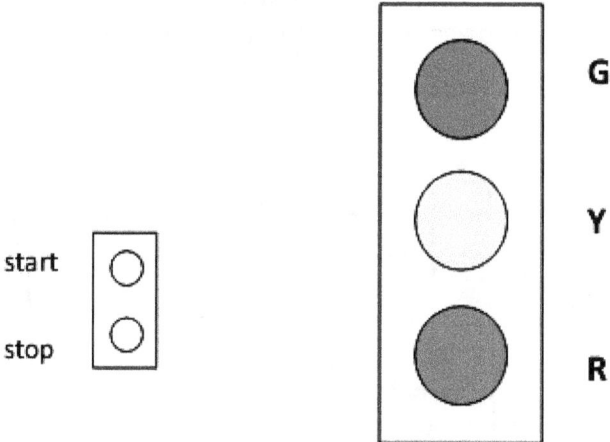

Problem 3:

By pressing the start push button pump1 and pump2 are triggered
simultaneously when the limit switch1 is activated pump1 is
switched off. When the limit switch3 is activated pump2 is switched
off. When limit switch1 and limit switch3 are both activated, valve1
and valve2 are opened simultaneously. Valve1 is also closed when
limit switch2 is deactivated. When limit switch4 is deactivated
valve2 closes. When limit switch2 and limit switch4 are both deactivated the mixer will start for 100 seconds and then valve3 will
open. When limit switch5 is deactivated, valve 3 closes and the process is terminated. This process cycle is repeated sequentially. By
pressing the stop button, the entire system resets and the start button must be pressed again to start this process cycle. Design a LAD to
control this system.

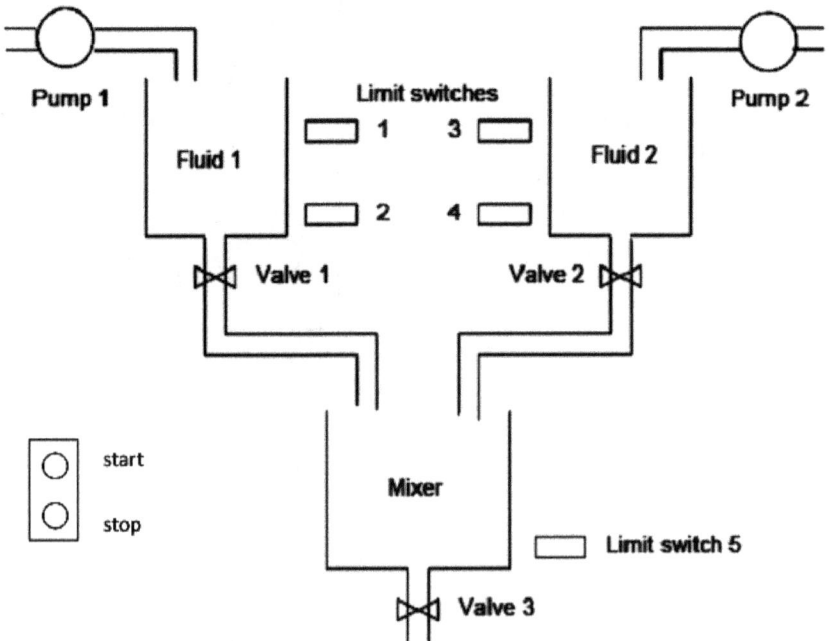

Problem 4:

When the start switch is turned on the system is ready to work. When the operator places a box at the top of the conveyor, (on the LS1) the motor runs and the conveyor stops when reaching the center point of the conveyor (on the LS2). Then the stamper goes down to the LSDN and places the stamp on the box. When this process ends, the stamp goes up to the LSUP and the conveyor moves to the other end. The motor stops after the box reaches the end of the conveyor (on the LS3). The system waits until the box is removed and the other box is placed at the top of the conveyor (on the LS1). If the start switch is turned off, the system cannot operate even if there is a box on the conveyor. The light bulb on the start switch indicates that the system is active. Up and Down lamps, indicate that the stamp is in the upper and lower position respectively. Design a LAD to control this system.

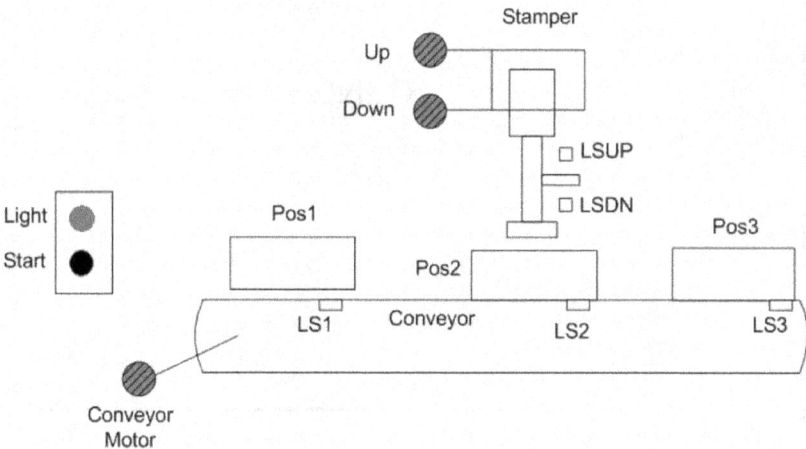

Problem 5:

A parking lot with a capacity of 100 cars is to be controlled by a PLC system. The S1 and S2 sensors are used to count the car in the Entrance and Exit. If the number of cars reaches 100, the Red Light bulb lights up and the Gate Arm closes. The Gate Arm will remain closed until one or more parking spaces are available. The Gate Arm is controlled by activating or deactivating the Gate Solenoid (GS). Design a LAD to control this system.

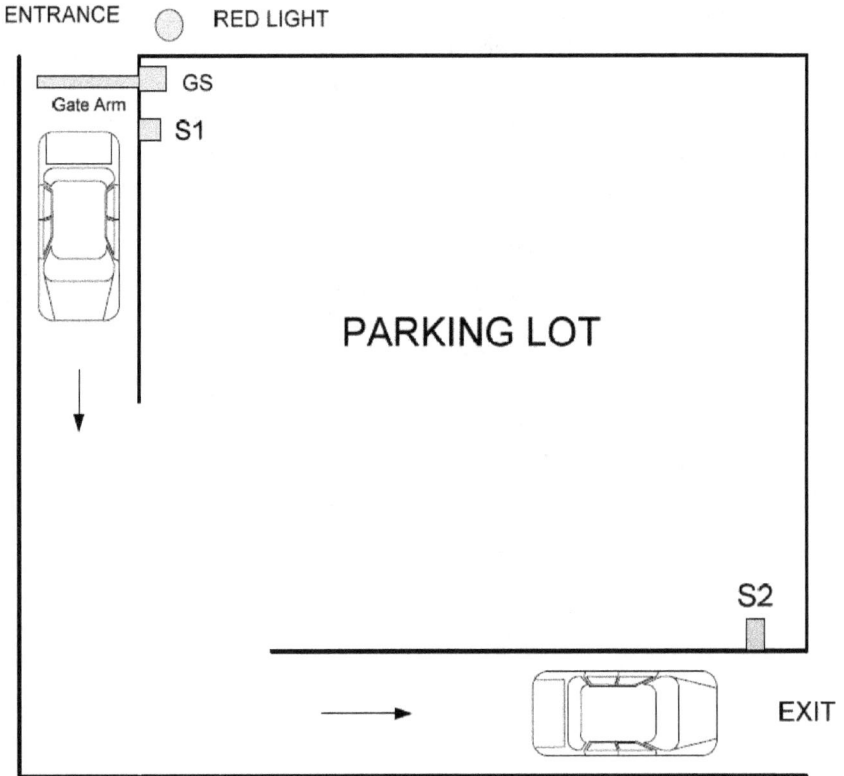

Problem 6:

By pressing the start button, the boiler starts and the M1 and M2 valves open immediately to allow hot water to flow into the radiators and hot water tank. The boiler is on until the boiler temperature sensor or room temperature sensor and hot water tank temperature sensor are activated. If room temperature sensor is activated the M1 valve is closed also if hot water tank temperature sensor is activated the M2 valve is closed. When the room temperature sensor is switched off, the M1 will open and the boiler will switched on. In addition, when the hot water tank temperature sensor is deactivated, the M2 valve opens and the boiler is turned on. By pressing the stop button, the boiler is turned off, the valves M1 and M2 are closed after 5 minutes, and the start button must be pressed again to start this process. Design a LAD to control this system.

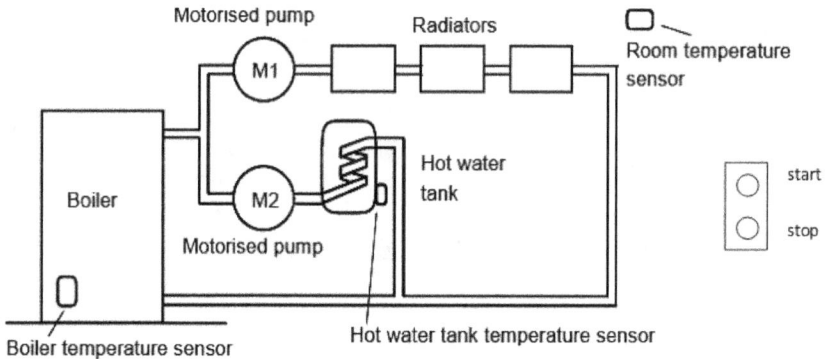

Motorised pump

Radiators

Room temperature sensor

Boiler

M1

M2

Hot water tank

Motorised pump

start

stop

Boiler temperature sensor

Hot water tank temperature sensor

Problem 7:

By pressing the start button, the exhaust fan starts and after 5 seconds if the flow sensor1 is switched off the exhaust fan turns off.
However, if the flow sensor1 is activated, the fresh air fan turns on
and after 5 seconds if the flow sensor2 is deactivated, the fresh air
fan and the exhaust fan both turn off. However, if the flow sensor2 is
activated the system will continue to operate. In this case, if either
the flow sensor1 or the flow sensor2 or both are deactivated the
fresh air fan and exhaust fan turn off after 5 seconds. By pressing the
stop button, both the fresh air fan and the exhaust fan are switched
off immediately. Design a LAD to control this system.

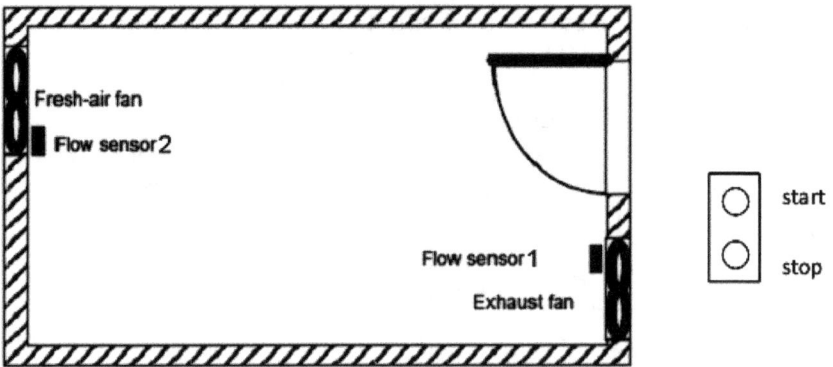

Problem 8:

By pressing the open button, the indicator lamp illuminates for 5 seconds before the door is opened, and the indicator lamp is turned on when the door opens. When the LS2 switch is activated, the door is stopped and the indicator lamp turns off. By pressing the close button, the indicator lamp is lit for 5 seconds before closing the door then the door is closed. During closing the door, the indicator lamp turns on. When the LS1 switch is activated, the door is stopped and the indicator lamp is turned off. A safety pressure bar is used to prevent damage to persons and objects when the door is closed and when the safety pressure bar is activated while the door is closed, the door is stopped and indicator lamp is turned off. Also by pressing the stop button, the door is stopped and the indicator lamp turned off. Design a LAD to control this system.

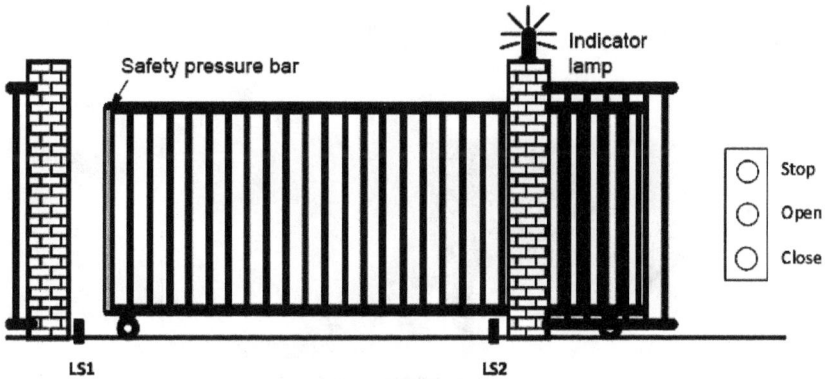

Problem 9:

By pressing the start button, the MV1 opens and the fluid fills the tank. At that moment the Mixer motor starts. When the liquid level crosses TBL2 and reaches TBL1, the MV1 will close and the Mixer motor will stop. The MV2 then opens and the fluid exits the tank. When the liquid level reaches TBL2, MV2 closes. This process is repeated three times and then the system is stopped. When the system is stopped, the Lamp is lit and the Buzzer is activated. After the system is stopped, the Buzzer will sound for 5 seconds and then switched off, but the lamp stays on until the reset button is pressed. If the stop button is pressed the system stops and the reset button should be pressed to restart before pressing the start button. Design a LAD to control this system.

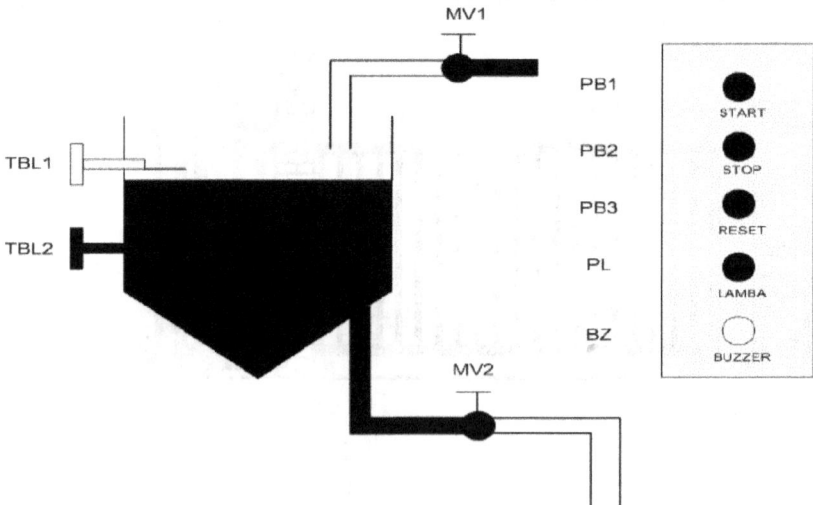

Problem 10:

The following system has two conveyors including Ball-Conveyor and Box-Conveyor. By pressing the start button, the Box-Conveyor starts moving. When the box stimulates the SE2 sensor, the Box-Conveyor stops and the Ball-Conveyor begins to move. The WS weight sensor measures the weight of the balls. If the ball weighs 500 g, the WS sensor is turned on. The Ball-Conveyor carries two different weights of balls (250 g, 500 g). If the ball weighs 500 g, the box is filled with 3 balls, otherwise the box is filled with 5 balls weighing 250 g. The SE1 sensor is used to count the balls. After the box is filled with balls, the Ball-Conveyor stops and the Box-Conveyor begins to move. The system can be reset by the stop button at any time. Design a LAD to control this system.

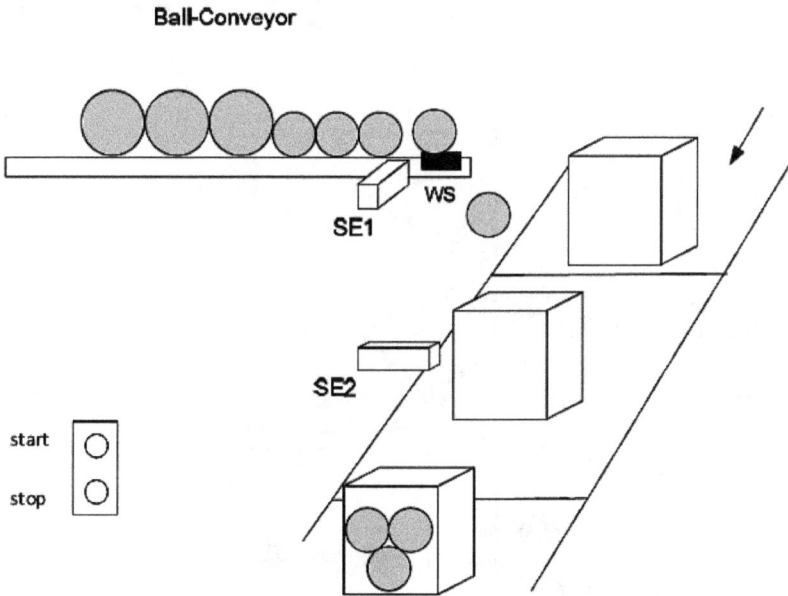

Ball-Conveyor

Problem 11:

An elevator system can move up and down with the up and down buttons. End positions are recognized by the limit switches (LS-Up and LS-Down). In the final positions, the elevator can only move in the opposite direction. The stop button stops the movement of the elevator. Elevator movement continues with manual control if the up or down push buttons are pressed for more than 2 seconds. If the emergency stop switch is turned on, the elevator will stop immediately and the elevator can only move with the release of the emergency stop (the up and down buttons do not work). A warning lamp illuminates as the elevator moves up or down. Design a LAD to control this system.

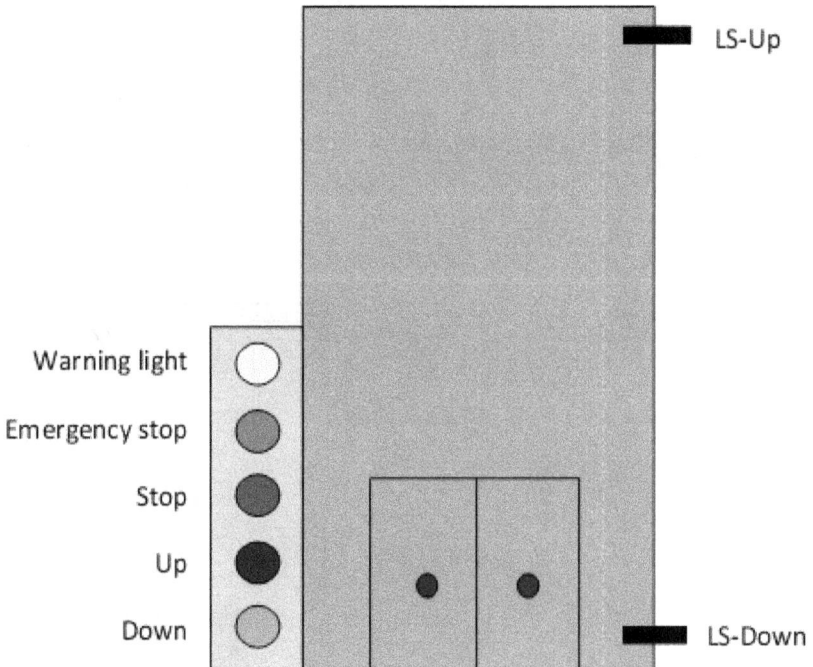

Problem 12:

The order of system performance is as follows:
- Start (NO) and stop (NC) push buttons are used for process start and stop.
-When the start push button is pressed the SOL A becomes energized and the tank starts to fill.
-When the tank is full, the empty sensor is closed and when the tanker is full, the full sensor is closed.
-SOL A becomes inactive when the full sensor closes.
-Then the mixer motor works for 3 min.
-When the mixer motor is stopped, SOL B is energized to empty the tank.
-When the tank is completely empty, the empty sensor opens and makes SOL B inactive.
- Pressing the stop button will stop the process.
- The start button is pressed to repeat the process.

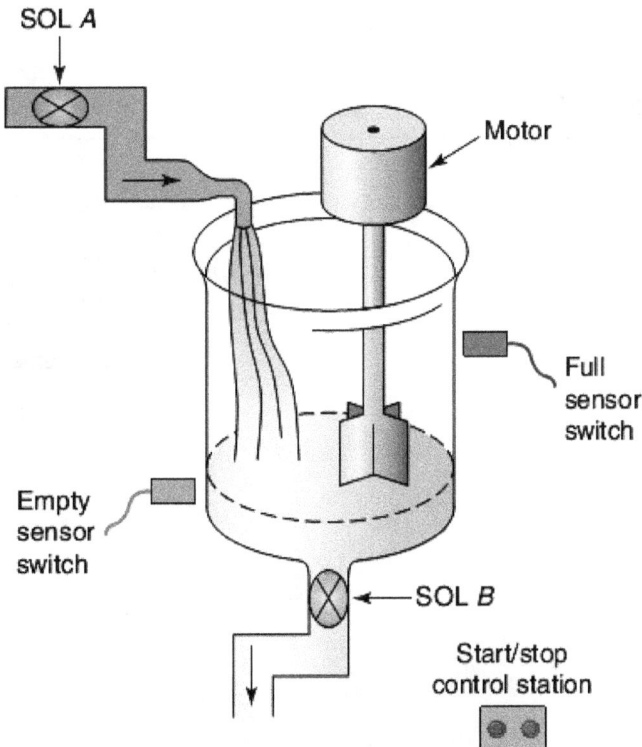

Problem 13:

The order of performance of the system is as follows:
- The box is in LS1 position (LS1 closed).
- By pressing the start button, the conveyor motor starts and the box moves to position A (LS1 opens).
- The conveyor moves the box to position A and then stops (the position is detected by eight off-to-on pulses from the encoder to up counter).
- After a 10 second delay, the conveyor starts to move and the box moves to LS2 and stops (LS2 closes).
- An emergency stop button is used to stop the process at any time.
- If the process is stopped by the emergency stop button, the timer and counter will reset automatically.

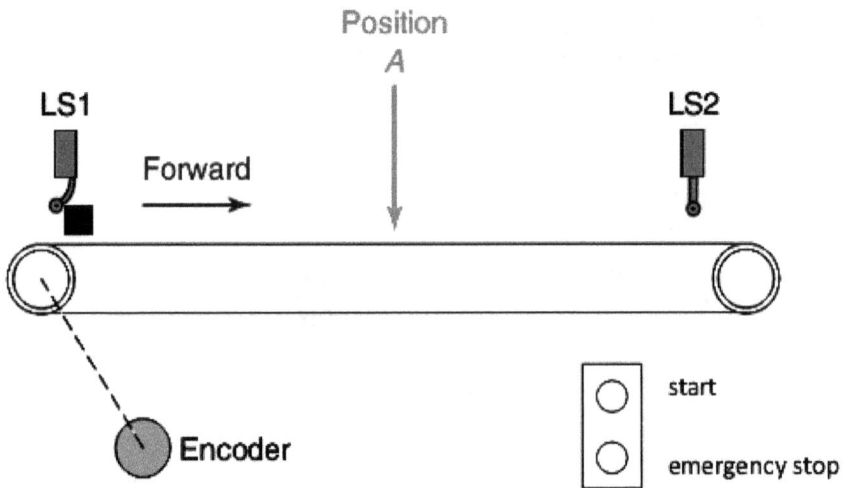

Problem 14:

The order of performance of the system is as follows:
- Start/stop push buttons are used to turn the conveyor motor on and off.
- A proximity switch is used to count pieces on the conveyor.
- When the count reaches 1000, the output of the counter activates solenoid gate and diverts the part to the inspection line.
- The solenoid gate is energized for 2 seconds to have enough time to move on the quality control line.
- After 2 seconds the gate returns to its normal position.
- The counter is reset and the counting is performed again.
- A reset push button is used to manually reset the counter.

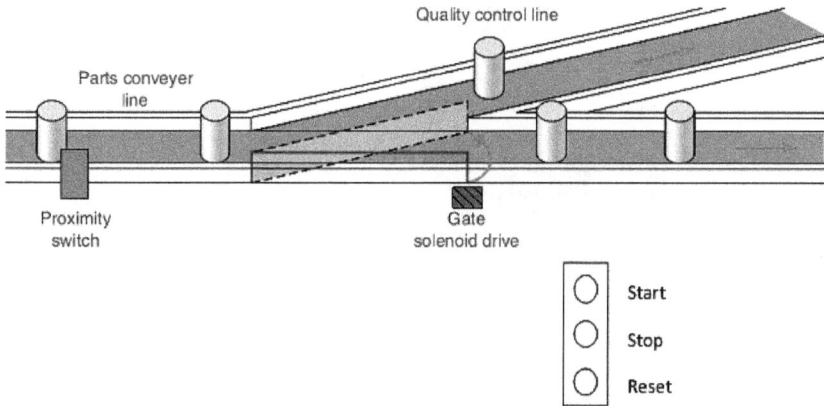

www.ingramcontent.com/pod-product-compliance
Lightning Source LLC
Chambersburg PA
CBHW071333210326
41597CB00015B/1441